D1175162

Unless Recalled Earlier

DATE DUE

Roaming in Wireless Networks

Shahid K. Siddiqui
Principal Consultant
Agilent Technologies
Kuala Lumpur, Malaysia

McGraw-Hill

New York   Chicago   San Francisco   Lisbon   London   Madrid
Mexico City   Milan   New Delhi   San Juan   Seoul
Singapore   Sydney   Toronto

The **McGraw·Hill** Companies

CIP Data is on file with the Library of Congress

TK
5103.2
.S54
2006

1 2 3 4 5 6 7 8 9 0 DOC/DOC 0 1 0 9 8 7 6 5

ISBN 0-07-145505-1

The sponsoring editor was Steve Chapman and the production supervisor was Richard C. Ruzycka. It was set in Century Schoolbook by International Typesetting and Composition. The art director for the cover was Anthony Landi.

Printed and bound by RR Donnelley.

This book is printed on recycled, acid-free paper containing a minimum of 50% recycled, de-inked fiber.

6565597

Dedicated to the loving memories of my mother

ABOUT THE AUTHOR

SHAHID K. SIDDIQUI is Principal Consultant working with Agilent Technologies in Malaysia. He has 20 years of experience in diverse areas of telecommunication including research and development, validation, test and measurement, and operations support systems. He works closely with wireless service providers delivering consulting and training. He has developed and taught technical training courses on digital switching, signaling, and wireless communication, and delivered numerous seminars.

Contents

Preface

Mobility is the key to the success of wireless networks. Roaming has extended the definition of mobility beyond the technology, network, and country boundaries. Is not it fascinating to make or receive calls anywhere in the world using the same phone and identity? International roaming is already proven to be one of the most popular features of today's wireless network. With the advent and widespread deployment of GSM technology, the mobile users have flexibility to use services in more than 500 networks. Inter-standard roaming has also made significant progress in recent years. Roaming capability in GPRS and 3G networks is progressively being implemented. The convergence of wireless mobility with Wi-Fi/WiMax is on card. The day is not far when a mobile user will be able to seamlessly roam anywhere regardless of the network, location, and device.

Interworking technology and operation and the business processes that enable roaming are complex. The main purpose of this book is to provide readers a comprehensive overview of roaming implementation, architecture, and protocols and its evolution from voice roaming in GSM to data roaming in GPRS and 3G networks. It can also be used as a guidebook to those who are responsible for roaming testing, maintenance, and management.

Chapter 1 introduces the key flavors of the roaming services to the readers. Chapter 2 provides an overview of Common Channel Signalling System number 7 (CCS7 or SS7). The CCS7 is the basis for the inter network communication between two wireless networks and it plays a key role in enabling roaming between two partner networks. Understanding of CCS7 is must to grasp the concept of roaming. Chapters 3, 4, and 5 provide an overview of GSM, GPRS and 3G networks, and the protocols. Understanding of radio and core network protocols is required to appreciate the inter network roaming transactions and call procedures. Chapter 6 focuses on the inter PLMN network infrastructure requirements for roaming and procedures for voice roaming in GSM network. Chapter 7 describes the additional requirements to enable data roaming in GPRS and 3G networks and the associated protocols. Chapter 8 discusses the issues related to implementation of roaming for the prepaid subscribers. It also describes the available alternatives to implement roaming for the prepaid subscribers. Chapter 9 focuses on the inter PLMN roaming tests, which are performed before launching roaming services in collaboration with a partner network. Chapter 10 discusses the issues and challenges to manage roaming services. It also describes the best practices to isolate and diagnose common roaming faults. The billing and settlement procedures for roaming services are quite different from the retail and wholesale billing for other services. Chapter 11 discusses the billing process and the format specified for the usage of services in foreign networks. To enhance the customer experience while roaming, the wireless service providers are constantly introducing new value-added services. Chapter 12 discusses few of the popular services and implementation. New radio access technologies such as WLAN and WiMAX offer new potential and opportunities. Chapter 13 discusses WLAN and PLMN convergence and possible roaming scenarios.

Shahid K. Siddiqui

Acknowledgments

First, I wish to thank Steve Chapman, Editorial Director, McGraw-Hill, for offering me an opportunity to write this book. A special thanks to Diana Mattingly, Acquisitions Coordinator, McGraw-Hill, for the guidance during acquisition and Samik Roy Chowdhury (Sam) and his competent staff members at International Typesetting and Composition (ITC) for providing editorial services and production of the book.

I am grateful to my colleague Gordon Law for reviewing few initial chapters and providing me useful inputs and much needed encouragement.

I am thankful to my current employer Agilent Technologies, Malaysia; past employers Hewlett Packard, India and Malaysia; and Centre for development of Telematics, India, for providing me an inspiring working environment to learn and practice telecommunication.

Finally, I like to express my sincere thanks to my wife Shazia, son Hamza, and daughter Yusra for the patience and support during my efforts to complete this project. They have surely missed many weekends and holidays.

Roaming and Wireless Networks

Roaming is one of the most popular features offered by wireless networks today. For mobile users, it offers the ability to use the mobile services outside their service provider's coverage area with the same phone. For service providers, roaming offers an opportunity to serve visitors from foreign networks as well as their own subscribers anywhere and anytime. It is also a most profitable revenue stream for the wireless service providers.

Roaming was introduced in the very first generation of cellular networks but was not available on a global basis till recent years. The early standards for cellular networks were focused in standardizing the Common Air Interface (CAI). There was not much work done in standardizing internetwork communication, resulting in a variety of vendor-dependent proprietary protocols. This means that roaming was possible only between two networks supplied by the same vendor. As the demand for roaming increased, the need for standards for communication between home and visited network was felt. The IS-41 standard was introduced as a standard protocol for internetwork communication to enable roaming in AMPS-based networks. Later, as part of GSM standardization, Mobile Application Part (MAP) was developed. Both IS-41 and GSM MAP were enhanced several times to ensure seamless roaming for the next generation of networks.

Today, with multimode mobile phones supporting GSM 900/1800/1900, it is possible to roam in a visited network with different radio frequencies. New UMTS phones are backward-compatible with GSM/GPRS networks. This allows 3G subscribers to roam in GSM/GPRS networks when they are outside 3G coverage. This is a very important feature, as initial deployment of 3G networks is unlikely to cover the entire nation because of cost constraints.

During the last few years, GPRS and 3G networks were deployed. Roaming in a GPRS/3G network is not an automatic extension of GSM voice roaming. The service providers are progressively building necessary infrastructure and services to offer true seamless global roaming as envisioned in 3G specifications. It may take a few years before GPRS/3G roaming can reach a level comparable to GSM roaming.

1.1   National and International Roaming

Roaming is the ability for a mobile subscriber to make/receive voice calls, send/receive data, and use other value-added services in a visited network, outside the geographical coverage area of the home network. The home and visited networks are referred to as the Home Public Land Mobile Network (HPLMN) and the Visited Public Land Mobile Network (VPLMN) respectively. In the case of international roaming, the HPLMN and the VPLMN belong to two different countries.

Not all wireless service providers offer their mobile services across national boundaries. This constraint may be because of licensing, technical, or commercial reasons. For example in India, a license to run mobile networks was initially based on "circles," each circle consisting of one or more states or provinces. In order to offer a nationwide service, a wireless service provider offers roaming within national boundaries. In the case of national roaming, the HPLMN and the VPLMN belong to the same country.

1.2   Interstandard Roaming

Interstandard, or cross-technology, roaming refers to roaming capabilities between two networks regardless of technology and standards. From business and user satisfaction points of view, interstandard roaming capabilities are required to further expand roaming services. The incompatibility in standards makes it difficult to enable roaming between networks. For example, second-generation networks are predominantly based on the GSM TDMA or the CDMA access technologies. GSM networks deploy GSM MAP and CDMA networks deploy IS-41 for internetwork communication. GSM uses the HLR for user authentication and CDMA uses HLR and AAA. Implementing roaming between these two islands of networks is not straightforward. Actually, the issues faced by the industry are more than technical. They include:

- Interoperability issues related to access technologies, handsets, smart cards
- Interoperability issues related to signaling protocols
- End-to-end testing
- Interoperator roaming agreements

- Exchange of usage/billing information and format
- Revenue assurance, e.g., prevention of fraud
- Security

In addition to conventional mobile networks, one has to consider the convergence of the technologies such as WLAN and WiMax.

Most of the interstandard roaming solutions available today are based on SIM roaming. This allows subscribers to use their own SIM and an intelligent handset. For example, if a GSM subscriber wishes to roam in a CDMA network, he/she rents a special phone, which accepts GSM SIM. This enables a roamer to retain personal information and MSISDN number. The network provider or roaming hub deploys a protocol converter (generally known as roaming gateway) to convert IS-41 signaling to GSM MAP and vice versa.

The other solution requires a dual CDMA/GSM handset. These handsets are commercially available but are expensive. These handsets also support dual slots for smart cards, i.e., SIM and RUIM.

One of the important requirements laid down by ITU-T for 3G networks is the capability of seamless roaming. The standardization and harmonization efforts on access technologies ensure that the 3G subscribers are able to roam in any network on a global basis. The 3G mobile phones also support WCDMA/TDMA/CDMA access technologies. This means that 3G subscribers can roam in a GSM or a CDMA network when outside 3G geographical coverage.

1.3   Prepaid and Postpaid Subscriber Roaming

International roaming allows a subscriber to access services virtually anywhere in the world. The visited network obviously needs to charge foreign subscribers for access time, transport and services. As the visited network is not in position to directly bill the roamers, it invoices their home network for the services usage. The home network then charge its own subscribers for the services used while roaming in a foreign network, using standard retail billing mechanism. This process works fine with postpaid subscribers as they are known to a service provider and bound by subscription agreement to pay for the services. The prepaid subscribers, on the other hand, pay for the services upfront. The service providers perform credit checks on each service initiation and decide if the service shall be provided or not. In case of roaming, the VPLMN, which processes and controls the calls/services initiated by a roamer has no information on available credit. To enable roaming for prepaid, the HPLMN must take control of all the services initiated by a prepaid subscriber in a foreign network. Therefore, roaming implementation

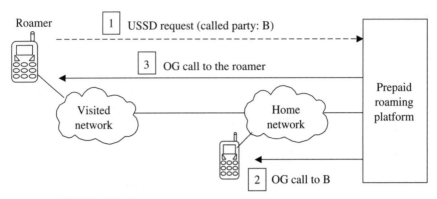

Figure 1-1 USSD callback.

for the prepaid subscribers is different from the implementation for their postpaid counterparts.

Today, most of the prepaid implementations are based on CAMEL or USSD callback. CAMEL is a network feature. CAMEL allows users to access services in a visited network transparently. Using CAMEL, the HPLMN has full control on the services used by a prepaid roamer in a visited network. Both the HPLMN and the VPLMN must be CAMEL-enabled to implement prepaid roaming services.

USSD callback allows a prepaid roamer to request to make a call in a foreign network by sending the called party number in a predefined USSD message. Figure 1-1 shows the USSD callback concept. The message is routed back to the prepaid system in the home network. The prepaid platform then performs the necessary credit checks and initiates two outgoing calls, i.e., one to the roamer and another call to the called party as requested by the roamer. The HPLMN monitors the call and may decide to disconnect after appropriate notification if the credit balance is running out.

1.4   Basic Structure of Roaming

In order to enable roaming, following basic structure should be in place:

1. *Inter-PLMN connection.* With reference to Figure 1-2, this connection consists of:
   (a) CCS7 links (for SCCP MAP traffic) between the VPLMN and the HPLMN. These links are required for information exchange between the HLR in the home network and the VLR in the visited network.
   (b) Interconnect links to transport circuit-switched voice and data between the HPLMN and the VPLMN.

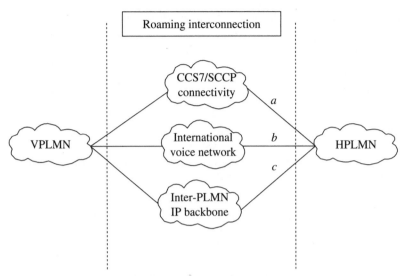

Figure 1-2 Inter-PLMN connection.

(c) Packet-switched interconnection to transport packet data between the HPLMN and the VPLMN.

Roaming in GSM networks requires (a) and (b) types of interconnection. GPRS roaming requires (a) and (c) types of interconnection. For 3G, all three types of interconnection are required.

2. *Agreement.* To allow a subscriber to roam and use services in a VPLMN, the two networks (the HPLMN and the VPLMN) must have a roaming agreement in place. The roaming agreement can be a bilateral agreement between two roaming partners or it can be an indirect relationship using a clearinghouse or a roaming broker. The roaming agreement covers several operational and business aspects including interconnection, problem resolution, tariff, pricing, and usage data format and exchange mechanism.

3. *Billing.* The VPLMN generates usage records for all the services used by a roamer while staying in the network. It then rates the usage records and raises the invoice to the roamer's HPLMN on the basis of the terms and conditions set in the roaming agreement. The VPLMN also transfers the detailed usage records of each individual roamer to the HPLMN in a specified format. The HPLMN settles the invoices with the VPLMN and charge its own subscriber for the service usage while roaming. The billing and settlement process between two operators can be direct or through a clearinghouse.

4. *Testing.* Interworking tests are performed before the roaming is commercially launched. This is required to ensure that the user can access

all the services provided by the roaming agreement. On-demand and periodic tests are also performed to ensure roaming availability in view of continuous changes in network and services.

1.5   Roaming Services

The service a roamer enjoys in a visited network depends on three factors: mobile station (MS) capabilities, the agreed list of services in the roaming agreement, and the subscription level.

Commercially available handsets generally support the following network capabilities:

- GSM
- GSM + GPRS
- GSM +GPRS +3G

A complete list of services available to a roamer in a visited network is given in Chapter 9.

2

CCS7 in Wireless Networks

Common Channel Signaling System no. 7 (CCS7) was initially designed for fixed line networks. As we will learn in the following chapters, CCS7 is also a basis for signaling traffic in the GSM core network and plays an important role in 3G networks after suitable adaptation. An understanding of CCS7 is required to grasp the signaling concepts in wireless networks. This chapter introduces the CCS7 network architecture, its layered protocol architecture, and the user parts. CCS7 is also commonly known as Signaling System 7 (SS7).

2.1 Signaling—An Introduction

By definition, signaling is the process of transferring information over a distance to control the setup, holding, charging, and releasing of connections in a communication network. In the past, several different types of signaling system were in use. Some examples of signaling used in core networks are: CCITT, R1, CCITT R2 (National network), CCITT C5, and CCITT C6 (International network).

Prior to CCS7, Channel-Associated Signaling (CAS) was used. In CAS, a dedicated signaling link is required for each speech channel. For example, if a 30-channel PCM is used to interconnect two telephony exchanges, the dedicated signaling channel for each bearer is multiplexed and carried in one of the channels, e.g., in time slot 16. This is not an efficient utilization of resources and is slow, resulting in long call setup time. With the advent of CCS7, a logically separate signaling network is established to transport the signaling information from a large number of bearers. For example, one 64-Kb/s signaling link can carry signaling information for the control of 4096 speech circuits. In addition to its economical use of PCM channels, CCS7 can support a wide range of services and more message types and is much faster. CCST is used both in national and international networks.

2.2 CCS7 Network Architecture

The CCS7 network is a logically separate network within a telecommunication network. It consists of signaling points or signaling nodes connected with the signaling links. The CCS7 network has four distinct signaling points.

Service signaling points (SSPs) are network nodes that generate signaling messages to transfer call- or transaction- (non-call-) related information between different CCS7 nodes. In wireline networks, a local switch may have SSP capabilities. In wireless networks, the BSCs and MSCs are the SSPs.

Signaling transfer points (STPs) are network nodes that relay signaling information from one signaling node to another.

A *combined SP/STP* is a node that has capabilities of both SP and STP; i.e., it can originate or accept CCS7 signaling messages as well as transfer messages from one SP to another SP.

Signaling control points (SCPs) are nodes that contain databases that enable enhanced services.

Signaling links interconnect two signaling points. A signaling linkset is made up of multiple signaling links. It is recommended to have at least two signaling links in a linkset for reliability purposes. A linkset can have a maximum of 32 links. A route is defined as a collection of links between originating and terminating SPs via intermediate nodes. There may be several routes that a message can traverse between the originating and terminating SP. These signaling routes are collectively called a signaling routeset.

Figure 2-1 shows a simplified CCS7 signaling network architecture. As we will learn later in this chapter, the CCS7 protocol has a built-in

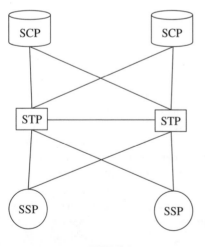

Figure 2-1 CCS7 signaling network architecture.

——— CCS7 links

error recovery mechanism to ensure reliable transfer of signaling messages. To take full advantage of the built-in recovery mechanism, the STPs and SCPs are generally provided in mated pairs. In addition, redundant links are provided to transfer the signaling messages using alternate routes in case of link failure.

CCS7 has a layered protocol architecture, as shown in Figure 2-2. The protocol stack consists of four levels. These levels are loosely related to Open System Interconnects (OSI) Layers 1 to 7. The lower three levels, referred to as the Message Transfer Part (MTP), provide a reliable service for routing messages through the CCS7 network.

The Signaling Data Link (referred to as MTP Level 1) corresponds to the Physical Layer of the OSI model. It defines the physical and electrical characteristics of the signaling link connecting two signaling nodes.

The Signaling Link (MTP Level 2) corresponds to the Layer 2 of the OSI model. It is responsible for error-free transmission of messages between two adjacent signaling nodes.

The Signaling Network (MTP Level 3) provides the functions related to message routing and network management.

MTP Levels 1, 2, and 3 together do not provide a complete set of functionalities as defined in OSI Layers 1 to 3. The Signaling Connection and Control Part (SCCP) offers enhancements to the MTP Level 3. The SCCP and MTP together are referred as the Network Service Part (NSP).

At Level 4, there are several user parts or application parts. The user parts use the transport capabilities of MTP or NSP. ISDN User Part (ISUP) provides for the control signaling needed to support ISDN calls. The Transaction Capabilities Application Part (TCAP) provides the control signaling to connect to centralized databases. The Mobile Application Part,

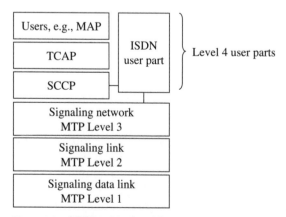

Figure 2-2   CCS7 protocol architecture.

which is the user of TCAP, provides the ability to support user mobility in wireless networks.

2.3 Message Transfer Part

2.3.1 MTP Level 1

The Signaling Data Link corresponds to the Physical Layer of the OSI model. It defines the physical and the electrical characteristics of the signaling link, connecting two signaling nodes. The Signaling Data Link is a bidirectional physical connection. The physical interfaces initially defined by ITU-T include:

- E1, 2.048 Mb/s, 64-Kb/s channel
- T1/DS1, 1.544 Mb/s, 56-Kb/s channel
- Other interfaces such as RS-232, RS449, DS-0, and V.35

In wireless networks CCST is also used to transport data such as SMS. SMS is a popular service and is growing at fast pace. To meet the increased demand, high speed links ($n \times 64$ or $n \times 56$ Kbps) are need. Furthermore, to exploit less expensive IP transport, new standards such as SIGTRAN are available to support CCS7 over IP.

2.3.2 MTP Level 2

The Signaling Link (MTP Level 2) corresponds to Layer 2 of the OSI model. It is responsible for error-free transmission of messages between two adjacent signaling nodes. The messages related to network management and maintenance and from the user parts are transferred between the nodes in data blocks called signal units (SUs). The functions of MTP Level 2 include:

- SU delimitation
- SU alignment
- Error detection and correction
- Signaling link error monitoring
- Initial alignment
- Flow control

To achieve these functions, MTP Level 2 adds overheads over the Level 3 information. As we will learn later in this section, SU delimitation is achieved by the use of flags at the beginning and the end of an SU, flow control is achieved by using forward and backward sequencing, and error control is achieved by means of cycle redundancy checks.

Signal unit format. There are three different types of signal unit that are transmitted via a signaling link:

Message signal unit (MSU) carries signaling information from the user parts for call control, network management, and maintenance. For example, ISUP MSUs carry call control messages for an ISDN call.

Link status signal unit (LSSU) carries link status control information.

Fill-in signal unit (FISU) is transmitted on the signaling link when there is no MSU or LSSU available to send.

The format of different types of SUs is illustrated in Figure 2-3, where:

F: Flag indicates the beginning and the end of a SU. Flag pattern = 01111110, bit stuffing is used to avoid occurrence of this pattern elsewhere in the SU.

CK: Cycle redundancy check is a 16-bit checksum of an SU.

BSN: Backward sequence number

BIB: Backward indicator bit

FSN: Forward sequence number

FIB: Forward indicator bit

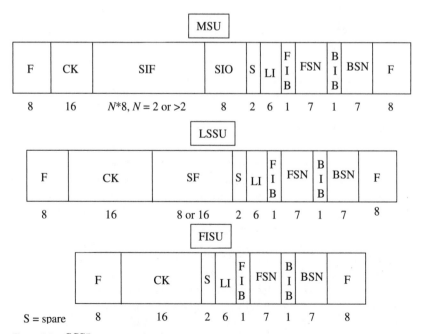

Figure 2-3   CCS7 message structure.

The BSN, BIB, FSN, and FIB fields are used for error and flow control. The flow control is based on a sliding window mechanism and error control is based on go-back-N automatic repeat request (ARQ) mechanism.

LI: Length indicator indicates the number of octets that follow the LI field and precede the CK field. Values of LI for different SUs are as follows:

- FISU: LI = 0
- LSSU: LI = 1 or 2
- MSU: 2 < LI < 63

SIO: Service information octet indicates the nature of a MSU. It consists of two subfields: sub-service field (SSF) and service indicator (SI). The service indicator indicates the user part, e.g., ISUP MSU, SCCP MSU, MTP SNM (Signaling Network Management) MSU etc. The sub-service field allows a distinction between the national and the international CCS7 networks.

SIF: Signaling information field contains Level 3 and Level 4 information, i.e., the routing label and user data. This will be discussed in more detail in the next section.

SF: Status field is part of LSSU. It indicates the status of the signaling link. The valid status indications are:

- Status indication O: out of alignment
- Status indication N: normal alignment
- Status indication E: emergency alignment
- Status indication OS: out of service
- Status indication PO: processor outage

S: Spare bits,

2.3.3   MTP Level 3

The Signaling Network (MTP Level 3) handles functions and procedures related to signaling message routing and network management. The MTP Level 2 is concerned with the individual signaling link, while MTP Level 3 functions relate to overall network aspects.

As Figure 2-4 illustrates, MTP Level 3 includes functions related to message handling and functions related to network management.

Message handling. The purpose of the message handling part in each signaling node is to transfer a signaling message originated by a particular user part to the same user part at the destination point indicated by the sender. The message handling comprises three functions: routing, discrimination, and distribution. Each signaling node uses the routing function to determine the outgoing signaling link to be used to forward

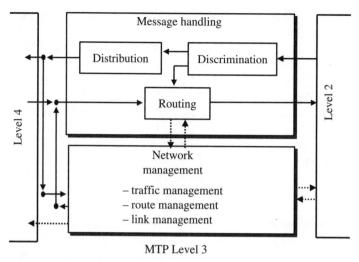

Figure 2-4   Signaling Network functions.

the message. The discrimination function is used to determine if a message received is destined to its node or is to be relayed to another node. The distribution function determines the user part to which a message should be delivered. The message handling decisions are based on the routing label contained in the SIF field. Figure 2-5 illustrates the contents of SIF in ISUP and SCCP MSUs. The routing label contains originating and destination point codes and a signaling link selection code.

Each signaling node in a CCS7 network is uniquely identified by its point code. The originating point code (OPC) indicates the source of the message, while the destination point code (DPC) identifies the destination

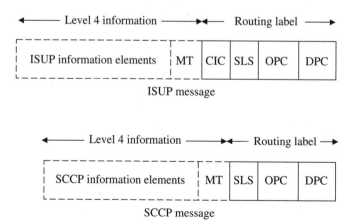

Figure 2-5   Signaling information field.

of the message. Figure 2-6 shows the format of point codes adopted by ANSI and ITU-T standards.

ITU-T point codes use 14 bits. Typically, a single number, e.g., 5555, is used to express a point code in a national network. For the international network, it is generally stated in terms of zone, area/network, and signaling point identification number, e.g., 6-0-0.

ANSI point codes use 24 bits (3 octets). It consists of network, cluster, and member octets, e.g., 22-7-0.

The Signaling link selection (SLS) is used to indicate the link in a linkset connecting two adjacent signaling points, over which a signaling message is to be routed. In practice, more than one link is used to connect signaling points. These links share the signaling load.

Signaling network management.   The signaling routeset availability objective set by the CCS7 specifications is very stringent. It calls for 99.9998% or better availability. This is equivalent to no more than 10 minutes unavailability per year for any route. This goal is achieved by monitoring the status of each link, with capability to reroute signaling traffic to overcome link degradation or outage. Unlike other systems where the network management part is outside the scope, the CCS7 includes this functionality to achieve desired routeset availability goals. The Signaling network management functions are divided into three categories:

- Signaling link management
- Signaling traffic management
- Signaling route management

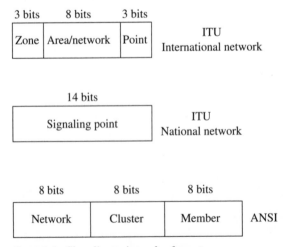

Figure 2-6   Signaling point codes format.

Signaling link management. The signaling link management function controls the links locally connected to an SP. This function ensures that predetermined linkset capabilities are maintained. It initiates action to activate additional links if required in the event of signaling link failure. The procedures supported by signaling link management functions are:

- Link activation
- Link restoration
- Link deactivation
- Linkset activation

Signaling traffic management. The signaling traffic management function is used to divert the signaling traffic from a link or a route to one or more different links or routes. This function, for example, could be used to divert the signaling traffic carried by the unavailable link to other available links. The redistribution of traffic may also be required to ease the congestion on one particular link or route. The signaling traffic management functions include the following procedures:

- Changeover
- Changeback
- Forced rerouting
- Controlled rerouting
- Management inhibiting
- MTP restart
- Signaling traffic flow control

Signaling route management. The signaling route management functions are used to exchange signaling route availability information between the signaling nodes in a CCS7 network. The signaling traffic management functions include the following procedures:

- Transfer prohibited
- Transfer restricted
- Transfer allowed
- Signaling routeset test
- Signaling routeset congestion and transfer control

2.4 ISDN User Part

Integrated Services Digital Network User Part (ISUP) provides the signaling functions required to control circuit-switched voice/data calls and supplementary services. ISUP is also used extensively in the GSM core network for controlling calls between MSCs and between the GMSCs and the external PSTN.

ISUP call control is achieved by the exchange of ISUP messages. These messages have a fixed structure consisting of a header to indicate message type, mandatory fixed parameters, and optional parameters. Figure 2-7 illustrates the ISUP message format.

Figure 2-8 shows an ISUP initial address message (IAM) message decode. The Mandatory fixed and variable parameters are as follows:

1. Message type
   - 01_{hex} for IAM
2. Nature of connection indicator
   - Satellite indicator: no satellite circuit
   - Continuity check indicator: continuity check not required
   - Echo suppressor indicator: O/G half-echo suppression not included
3. Calling party category
   - Ordinary calling subscriber
4. Transmission medium requirement
   - Transmission medium requirement: 3.1-kHz audio
5. Called party number

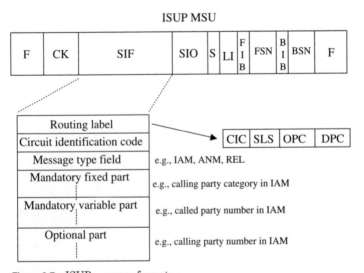

Figure 2-7   ISUP message format.

Blue Book ISUP (ISUP) Initial address (IAM)
----0101 Service indicator   ISDN User Part
--00---- Subservice: Priority   Spare/priority 0 (U.S.A. only)
10------ Subservice: Network indicator   National message
******** Destination point code   81xx
******** Originating point code   82xx
******** Signaling link selection 12
******** Circuit identity code   254 (PCM: 7 Channel:30)
0000---- Spare
00000001 Message type   0x1
------00 Satellite indicator   No satellite circuit
----00-- Continuity check indicator   Continuity Check not required
---0---- Echo suppressor indicator   O/G half echo suppression not included
000----- Spare
-------0 National/international indicator   Treat as a national call
-----00- End-to-end method indicator   No end-to-end method available
----0--- Interworking indicator   No interworking encountered
---0---- End-to-end information indicator   No end-to-end info available
--1----- ISDN User Part indicator   ISDN-UP used all the way
01------ ISDN-UP preference indicator   ISDN-UP not required all the way
-------0 ISDN access indicator   Originating access non-ISDN
-----00- SCCP method indicator   No indication
00000--- Spare
00001010 Calling party's category Ordinary calling subscriber
00000011 Transmission medium requirement   3.1-kHz audio
00000010 Pointer to called party number 2
00001010 Pointer to optional parameter   10
Called party number
00001000 Parameter length   8
-0000100 Nature of address   International number
0------- Odd/even indicator   Even number of address signals
----0000 Spare
-001---- Numbering plan indicator   ISDN numbering plan
(E.164/E.163)
0------- Internal network no. indicator   Routing to INN allowed
> ******** Called address signals   009512423xxF
Calling party number
00001010 Parameter name   Calling party number
00000111 Parameter length   7
-0000011 Nature of address   National (significant) number
1------- Odd/even indicator   Odd number of address signals
------11 Screening indicator   Network provided
----00-- Presentation restoration indicator   Presentation allowed

Figure 2-8 ISUP IAM protocol decode.

The rest of the parameters are optional. The IAM message is the longest ISUP message. It may contain up to 29 optional parameters.

Table 2-1 lists the ISUP messages and opcodes. There are 49 defined ISUP messages.

Figure 2-9 shows a basic call setup initiated by a fixed line subscriber to a mobile subscriber. To make the example simple, the signaling message flow within the GSM network is not shown.

1. On receiving a SETUP message from one of its subscribers, indicating origination and dialed digits, the local exchange analyzes the called party number and, on realizing that the call is to be routed to another exchange, uses the built-in SSP functionality to build an IAM message. This message contains all the necessary information that is required to route the call to the destination exchange.

2. An intermediate exchange, on receipt of the IAM, analyzes the destination address and other routing information and sends the IAM message to a succeeding exchange.

3. On receiving an IAM message, the GMSC (destination, in this example) uses the GSM procedures to locate the mobile subscriber and notify it of the incoming call.

4. The GMSC sends the ACM message back to the originating exchange via the intermediate nodes to indicate that the complete address of the called party has been received.

5. On receiving the ACM, the originating exchange passes an ALERTING message to the calling party.

6. On answer from the called mobile subscriber, the GMSC sends an ANM message to the originating exchange via the intermediate nodes.

7. The originating exchange sends a CONNECT message to the calling party to complete the call setup.

8. In the example shown in Figure 2-9, the calling party initiated the call release by sending a DISCONNECT message to the originating exchange.

9. The originating exchange then sends the REL message to the intermediate node and returns a RELEASE message to the calling party.

10. The intermediate node, on receiving the REL, returns an RLC to the originating exchange and forwards the REL to the destination exchange.

11. The GMSC, on receiving the REL, sends a DISCONNECT message to the called party and returns an RLC message back to the intermediate node.

TABLE 2-1   List of ISUP Messages

Mnemonics	Opcode (hex)	Message name
ACM	06	Address complete
ANM	09	Answer
BLO	13	Blocking
BLA	15	Blocking acknowledgment
CMC	1D	Call modification completed
CMRJ	1E	Call modification reject
CMR	1C	Call modification request
CPG	2C	Call progress
CRG	31	Charge information
CGB	18	Circuit group blocking
CGBA	1A	Circuit group blocking acknowledgment
GRS	17	Circuit group reset
GRA	29	Circuit group reset acknowledgement
CGU	19	Circuit group unblocking
CGUA	1B	Circuit group unblocking acknowledgment
CQM	2A	Circuit query
CQR	2B	Circuit query response
CVR	EB	Circuit validation response
CVT	EC	Circuit validation test
CSVR	25	Closed user group selection and validation request
CSVS	26	Closed user group selection and validation response
CNF	2F	Confusion
CON	07	Connect
COT	05	Continuity
CCR	11	Continuity check request
DRS	27	Delayed release
EXM	ED	Exit
FAA	20	Facility accepted
FAD	22	Facility deactivated
FAI	23	Facility information
FRJ	21	Facility reject
FAR	1F	Facility request
FOT	08	Forward transfer
INF	04	Information
INR	03	Information request
IAM	01	Initial address message
LPA	24	Loopback acknowledgment
OLM	30	Overload
PAN	28	Pass along
REL	0C	Release
RLC	10	Release complete
RSC	12	Reset circuit
RES	0E	Resume
SAM	02	Subsequent address message
SUS	0D	Suspend
UBL	14	Unblocking
UBA	16	Unblocking acknowledgment
UCIC	2E	Unequipped circuit identification code
USR	2D	User-to-user information

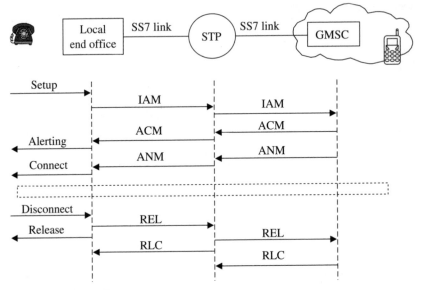

Figure 2-9   ISUP call setup.

2.5   Signaling Connection and Control Part

The SCCP supplements the MTP transport capabilities to provide enhanced connectionless and connection-oriented network services. Together with the MTP, it provides the capabilities corresponding to Layers 1 to 3 of the OSI model. The combined MTP and SCCP services are called the Network Service Part (NSP).

The SCCP structure is illustrated in Figure 2-10. As shown, it consists of four functional blocks.

1. SCCP connection-oriented (CO) control function handles the establishment, release, and supervision of the data transfer on logical signaling connections.

2. SCCP connectionless (CL) control provides the connectionless transfer of data units.

3. SCCP management handles the status information of the SCCP network.

4. SCCP routing handles the routing of SCCP messages. This includes routing based on global title and distribution of messages based on the subsystem number.

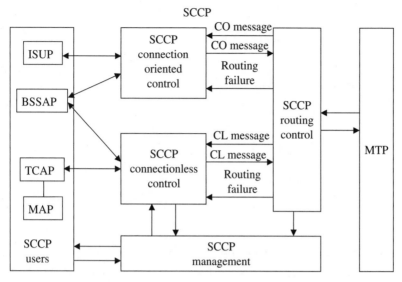

Figure 2-10   SCCP overview.

Four classes of network transport services are provided by the SCCP;

Class 0: Basic sequenced connectionless

Class 1: Sequenced connectionless

Class 2: Basic connection-oriented

Class 3: Flow control connection-oriented

2.5.1 Connectionless signaling

In both Class 0 and Class 1 connectionless services, the messages between SCCP users are transferred without establishing a logical connection. Each message is sent independently of the previously sent message. The SCCP user data is sent in a Unit Data (UDT) message. The difference between Class 0 and Class 1 is that Class 1 tries to offer (not guaranteed) in-sequence delivery by setting up the same SLS code for all the messages in a transaction. Figure 2-11 illustrates the data transfer between two SCCP users using SCCP functions at SSP-1 and SSP-2. The UDT contains the calling party (cgPA) and called party (cdPA) address, which identify the destination and origin of the message. Note that the UDT messages can take any available signaling path to reach the destination.

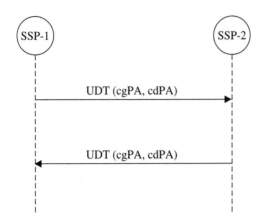

Figure 2-11   Connectionless services.

2.5.2   Connection-oriented signaling

In connection-oriented service, a logical connection is established between the SCCP users before the data transfer takes place. The logical connection establishment and subsequent data transfer procedure is shown in Figure 2-12.

1. On request from an SCCP user for a connection-oriented data service, the originating SSP-1 sends a CR (connection request) message to the SCCP located in SSP-2. In addition, to set up information—i.e., calling and called part address—SSP-1 also adds a source local reference (SLR = xx in the example shown in Figure 2-12).

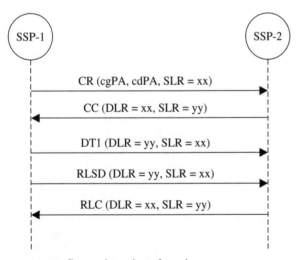

Figure 2-12   Connection-oriented service.

2. SSP-2, on receiving the message, returns a connection confirmed (CC) message to SSP-1. The CC message contains a destination local reference (DLR), which is set equal to the SLR value received in the CR message. It also adds its own SLR value (yy in this case).

3. With known values of an SLR and a DLR, a logical connection is now established. The user data can be exchanged between SSP-1 and SSP-2 using this logical connection. Each subsequent message from SSP-1 will have SLR = xx and DLR = yy. All the messages from SSP-2 will have SLR = yy and DLR = xx.

The user data is sent in Data Form 1 (DT1) for Class 2 and in DT2 for Class 3 of connection-oriented services. The Class 3 service provides flow control. This is achieved by assigning sequence numbers to each message. The monitoring capabilities in SCCP ensure in-sequence delivery and notification to SCCP users in case of message loss.

2.5.3   SCCP message format

The structure of SCCP message is shown in Figure 2-13. The SCCP messages are transferred between nodes in the Level 3 MSU. The SCCP MSU is identified by the SIO value, which is set to 03_{hex}. As shown in the figure, the SCCP message is of variable length. It consists of a routing label, message type, and a few mandatory and optional information elements.

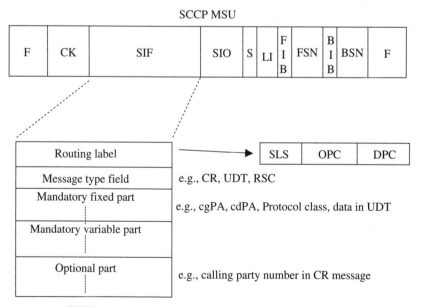

Figure 2-13   SCCP message format.

TABLE 2-2   SCCP Messages

Mnemonics	Message type	Opcode (hex)	0	1	2	3
					Protocol class	
CR	Connection request	01			√	√
CC	Connection confirm	02			√	√
CREF	Connection refused	03			√	√
RLSD	Released	04			√	√
RLC	Release complete	05			√	√
DT1	Data Form 1	06			√	
DT2	Data Form 2	07				√
AK	Data acknowledgment	08				√
UDT	Unitdata	09	√	√		
UDTS	Unitdata service	0A	√	√		
ED	Expedited data	0B				√
EA	Expedited data acknowledgment	0C				√
RSR	Reset request	0D				√
RSC	Reset confirm	0E				√
ERR	Error	0F			√	√
IT	Inactivity test	10			√	√

Table 2-2 lists the SCCP messages and the name of the protocol class that each message belongs to.

2.5.4   SCCP routing control

The purpose of SCCP is to enable end-to-end routing. This is intended to enhance MTP Level 3 point-to-point routing capabilities using point codes. In the case of SCCP, the routing is based on any combination of following elements:

- Point codes
- Calling and called party number
- Subsystem number

The calling and called party address information elements in an SCCP message contain one octet to indicate the address type and a variable number of octets containing the actual address. Two basic address types used are:

1. *Global title.* A global title (GT) is a regular directory number that does not contain the exact information to enable routing in a signaling

network. An SCCP translation function is required to derive routing information on each node.

2. Destination point code and subsystem number. The subsystem number (SSN) identifies an SCCP user function, e.g. VLR or MSC. Table 2-3 lists the defined SSN values and subsystem names. The DPC and the SSN combination allow direct routing by the SCCP and MTP without any translation required.

Figure 2-14 shows protocol decodes for an SCCP UDT message. The routing is based on global title and SSN is included.

2.5.5   SCCP management

SCCP provides its own management function. It is mainly intended to handle the status information of the SCCP network. This function also includes dynamic updating of routing table, based on the availability of subsystems (e.g., HLR or MSC). SCCP management messages are sent in the data part of UDT messages. The SCCP management function supports the following message types.

Subsystem status test (SST).   This message is used to probe a subsystem that has been reported as not available previously.

Subsystem prohibited (SSP).   This message indicates that a subsystem has been taken out of service.

Subsystem allowed (SSA).   This message indicates that a previously unavailable subsystem is now available.

TABLE 2-3   SSN Values

SSN (hex)	Subsystem
0	SSN not known
1	SCCP management
2	Reserved
3	ISUP
4	OMAP
5	MAP
6	HLR
7	VLR
8	MSC
9	EIR
0A	AUC
0B	SMSC
FE	BSSAP

BSN: 59 BIB: 0 FSN: 119 FIB: 1 LI: 63
SI: SCCP SSF: NN DPC: xxx OPC: yyy SLS: 14
MT: UDT
Protocol class: Class 0
Message handling: Return message on error
Pointer to called address: 3 octets
Pointer to calling address: 14 octets
Pointer to data: 25 octets
Called party address length: 11 octets
Routing indicator: Routing based on global title
Global title indicator: Transaction type, numbering plan, encoding scheme,
address indicator
SSN indicator: Address contains a subsystem number
Point code indicator: Address does not contain a signaling point code
Subsystem number: HLR
Translation type: 0
Encoding scheme: BCD, even number of digits
Numbering plan: ISDN/telephony
Nature of address indicator: International number
Address information: 886xxxxxxxxxh
Calling party address length: 11 octets
Routing indicator: Routing based on global title
Global title indicator: Transaction type, numbering plan, encoding scheme,
address indicator
SSN indicator: Address contains a subsystem number
Point code indicator: Address does not contain a signaling point code
Subsystem number: MSC
Translation type: 0
Encoding scheme: BCD, even number of digits
Numbering plan: ISDN/Telephony
Nature of address indicator: International number
Address information: 886yyyyyyyyyh
Data length: 39 octets

Figure 2-14   Partial decode of an SCCP UDT message.

2.6   Transaction Capabilities Application Part

In modern fixed and wireless networks, unlike the earlier versions, not all the network elements are switches. For example, in GSM, several databases are used that have no switching or routing capabilities of their own. The Transaction Capabilities Application Part (TCAP) provides a mechanism to establish non-circuit-related communication between any two nodes (as long as the nodes support MTP L1-3 and SCCP) and to exchange operation and replies using dialogues. In most of the applications today, TCAP is used to access remote databases such as the HLR or to invoke actions at remote network entities.

Figure 2-15 shows the TCAP in the CCS7 Layer and its relationship with the OSI reference model.

2.6.1   Structure of TCAP

As shown in Figure 2-15, TCAP is structured into two sublayers;

- Component sublayer
- Transaction sublayer

Component sublayer.   The component sublayer gets the information components from the TC users/applications. The TC user expects that these components will be delivered to the remote entity in sequence. The information components are basically the primitives and parameters necessary to invoke an operation at the remote entity.

The following five types of components are defined:

- Invoke
- Return result (not last)
- Return result (last)
- Return error
- Reject

The invoke component is used to request that an operation be performed at the receiver end. The invoke component indicates the operation type, using operation code. The operation codes are specific to a TC user.

The receiving entity, on receiving the invoke component with a specific operation code, performs the operation and returns the outcome of the operation in the return result component. It may be possible that one return result component may not be able to convey the result because of the

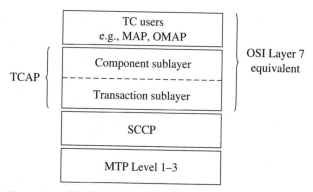

Figure 2-15   TCAP position in CCS7 protocol stack.

limited capacity offered by the UDT message. In such cases, the result data is segmented and transferred in the return result (not last) component. The last segment is transferred using the return result (last) component.

The return error component is used to report an unsuccessful operation. This does not necessarily mean a failure or fault. For example, if an entity invokes an operation in an MSC that results in paging a mobile station (MS), and the MS does not respond, then the return error component will indicate an unsuccessful operation.

The reject component reports the receipt and the rejection of an incorrect component. The reject component also reports if the application is unable to process a component because of problems.

Transaction sublayer.   The transaction sublayer is responsible for managing the exchange of MSUs containing components between two entities. It provides the necessary information to the signaling point to route the components to the remote entity.

There are five transaction sublayer message types supported:

- Begin
- Continue
- End
- Unidirectional
- Abort

The 'Begin' message is used to open a dialogue with a remote peer transaction sublayer. The begin message may include one or more components. The dialogue is identified by an originating transaction ID (OTID).

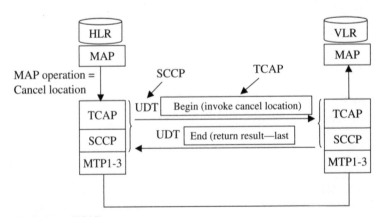

Figure 2-16   TCAP messages.

The 'Continue' message is used to transport additional information following a 'Begin' message. The first 'Continue' message in response to a begin message confirms the acceptance of application context and protocol. It comprises both OTID and terminating transaction ID (TTID). The 'Continue' message may have one or more components.

SI: SCCP SSF: NN   DPC: www OPC: zzz SLS: 10
MT: UDT
  Called party address length: 10 octets
    Subsystem number: VLR   visited location register
    Translation type: Translation type not used
    Nature of address indicator: International number
    Address information: 66xxxxxxxxh
  Calling party address length: 11 octets
    Subsystem number: HLR   home location register
    Translation type: Translation type not used
    Nature of address indicator: International number
    Address information: 601yyyyyyyh
  MT: Begin
  Originating transaction ID tag
    Transaction ID: 3a415d0ah
  Invoke
  Invoke ID tag
    Invoke ID: 1
  Operation code tag: Local operation code
    Operation: Cancel location
  IMSI tag
    MCC figits: 502 MNC digits: 1x MSIN digits: 122xxxxxxx

- -

SI: SCCP SSF: IN   DPC: --- OPC: ---- SLS: 15
MT: UDT
  Called party address length: 11 octets
    Subsystem number: HLR   home location register
    Translation type: Translation type not used
    Nature of address indicator: International number
    Address information: 601yyyyyyyh
  Calling party address length: 10 octets
    Subsystem number: VLR   visited location register
    Translation type: Translation type not used
    Nature of address indicator: International number
    Address information: 66xxxxxxxxh
  MT: End
  Destination transaction ID tag
    Transaction ID: 3a415d0ah
  Return result (last)
  Invoke ID tag

Figure 2-17 TCAP transaction and component sublayers.

The 'End' message is used to terminate a transaction. The 'End' message may have one or more components.

The unidirectional message is used for unstructured dialogue.

The unidirectional 'Abort' message is used to terminate a dialogue any time an error occurs or a requested operation cannot be processed.

Figure 2-16 shows that the MAP entity in an HLR invokes the cancel location operation in the remote entity, i.e., the VLR.

Figure 2-17 shows the protocol decode of the messages that flow between the HLR and the VLR to invoke a MAP cancel location operation in the VLR. Note that the MAP message is carried over as an invoke component in a TCAP dialogue initiated by the 'Begin' message and identified by a transaction ID 3a415d0a hex.

Bibliography

ITU-T Q.701, Functional Description of the Message Transfer Part of Signaling System No. 7.
ITU-T Q.702, Signaling Data Link.
ITU-T Q.703, Signaling Link.
ITU-T Q.704, Signaling Network Functions and Messages.
ITU-T Q.761, ISDN User Part of Signaling System No.7.
ITU-T Q.762, ISDN User Part. General Functions of Messages and Signals.
ITU-T Q.763, ISDN User Part. Formats and Codes.
ITU-T Q.764, ISDN User Part. Signaling Procedures.
ITU-T Q.711, Functional Description Signaling Connection Control Part.
ITU-T Q.771 Transaction Capabilities Application Part. Functional Description of Transaction capabilities.

Chapter

3

Global System for Mobile
Communication (GSM)

3.1 Brief History of Early Cellular Networks

Cellular telephone system, though with limited functionalities, features, and scale of deployment, existed as early as the 1920s. The commercial cellular networks, as we know them today, started in the late 1970s. The growth in mobile communication from then onward is amazing. Within 30 years of introduction, cellular telephony is now so popular that many will find it difficult to imagine life without it. The initial deployments of cellular networks were based on different variations of analog technologies and standards. These included:

- Advanced Mobile Phone Service (AMPS) was initially a United States standard but was later adopted by many other countries such as Australia, South Korea, Singapore, and Brazil. It operates in the 800-MHz band. Later Narrowband AMPS (NAMPS) was introduced by Motorola and adopted by operators in the United States, Russia, and other countries. NAMPS was developed as an interim technology between first and second generations of mobile networks. NAMPS, like AMPS, is based on analog technology. The significant difference lies in its voice channel bandwidth, i.e., 10 kHz instead of 30 kHz in AMPS.

- Total Access Communication System (TACS), a variation of AMPS, was initially introduced in the United Kingdom and later was deployed in many other countries such as Italy, Spain, and the United Arab Emirates. It operates in the 900-MHz band. A variation of TACS, known as JTACS, was later deployed in Japan.

- Nordic Mobile Telephone System (NMT) was initially introduced in the Scandinavian countries. NMT operates in the 450- and 900-MHz bands.

The NMT-450 and NMT-900 systems were deployed later in many countries in Asia and Europe and in Australia.

In addition, there were a few other limited implementations of cellular networks such as NETZ-C, which was deployed in Germany, Portugal, and South Africa.

3.1.1 Limitations of early cellular technologies

Early deployed cellular technologies caught the imagination of the users; they were a great success and registered phenomenal growth. However, they were limited by many factors. A few significant limitations of early cellular technologies are as follows:

- Restricted spectrum, limited capacity
- Cost of ownership
- Limited roaming
- Low mobility and cost of mobile phones
- Inherent speech quality issues
- Lack of internetwork standardization
- Compatibility issues with ISDN

3.1.2 Roaming and early cellular networks

The early standards for cellular networks were focused on standardizing the air interface, i.e., the Common Air Interface (CAI). There was not much work done in standardizing internetwork communication. The result was a variety of vendor-dependent proprietary protocols. This means the roaming was possible only between two networks supplied by the same vendor. This was surely a serious limitation when it came to roaming. As the demand for roaming increased, the need for a standard for communication between the home and visited networks was felt. The IS-41 standard was introduced as a standard protocol for internetwork communication to enable roaming in AMPS-based networks. Later, as part of GSM standardization, Mobile Application Part (MAP) was developed. Both IS-41 and GSM MAP were enhanced several times to ensure seamless roaming for the next-generation networks.

3.2 GSM Overview

In 1982, the Conference of European Posts and Telegraphs (CEPT) formed a study group to define and develop a pan-European standard for a mobile telephone system. This group was given the name Groupe

Special Mobile (GSM). The main task of this group was to propose a system to overcome inherent issues faced by the analog system existing at that time. The new system had to meet criteria defined by CEPT as given below:

- Spectrum efficiency
- Support for international roaming
- Lower cost of mobiles, infrastructure, and services
- Superior speech quality
- Support for a range of new services
- Compatibility with ISDN

Later, the study group was transferred to the European Telecommunication Standard Institute (ETSI), which released phase 1 of the GSM specification in 1990. The term GSM now means Global System for Mobile Communication. The GSM standard, which was initially developed for Europe, has been embraced worldwide. The standard has been evolving since then to meet demands of next generation networks.

GSM is feature rich. It includes automatic roaming, full voice and data services, excellent speech quality, and a wide range of supplementary services.

3.3   GSM Offered Services

GSM offers a wide range of services, including telephony, emergency calling, data up to 14.4 Kb/s, fax up to 9.6 Kb/s, SMS, and others. In addition, it also offers a rich set of supplementary services. According to ITU specifications, the telecommunication services are categorized into three different types, i.e., bearer services, teleservices, and supplementary services.

Bearer services are telecommunication services providing the capability of transmission medium between access points. These services are characterized by a set of low layer attributes.

Teleservices are telecommunication services providing complete capability, including terminal equipment functions. Teleservices are characterized by a set of low layer attributes, a set of high layer attributes, and operational and commercial attributes.

A supplementary service modifies or supplements a basic telecommunication service. It is offered together with or in association with a basic telecommunication service.

The following sections list the bearer services, teleservices, and supplementary services offered by GSM.

3.3.1 Bearer service

- Asynchronous data
- Synchronous data
- Asynchronous PAD (packet switched, packet assembler/disassembler)
- HSCSD—asymmetric
- HSCSD—symmetric

3.3.2 Teleservices

- Telephony (speech)
- Emergency calls (speech)
- Short message services
- Alternate speech and Group 3 fax
- Automatic Group 3 fax

3.3.3 Supplementary services

- Call forwarding
- Call barring
- Calling/connected line identity presentation (CLIP)
- Calling/connected line identity restriction (CLIR)
- Call waiting (CW)
- Call hold (CH)
- Multiparty communication—closed user group (CUG)
- Advice of charge (AoC)
- Unstructured supplementary service data (USSD)
- Operator-determined barring (ODB)

3.4 System Architecture

A GSM network is a most sophisticated and complex wireless network. It was designed for a purpose from scratch and is neither based on nor compatible with any predecessor technologies. In fact, it is the basis for future wireless networks such as GPRS, EDGE, and 3G.

A GSM network comprises several distinct functional entities:

- Mobile station
- Base station subsystem (BSS)
- Network switching system (NSS)
- Operation and support subsystem (OMC)

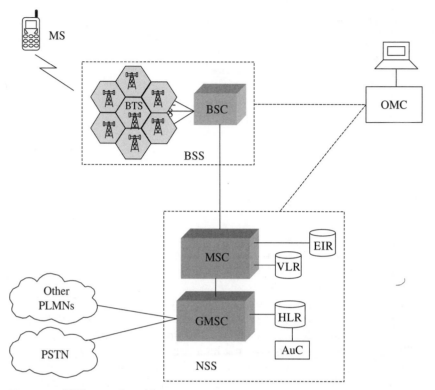

Figure 3-1 GSM network architecture.

Figure 3-1 shows a typical GSM network and its components.

The mobile stations talk to the BSS over the RF interface. The BSS consists of the base transceiver station (BTS) and base station controller (BSC). The BSS is responsible for management of connections on the radio path and handovers. The NSS consists of a mobile switching center (MSC) and databases. The NSS is responsible for management of subscriber mobility, interfacing with PSTN/other PLMNs, and end-to-end call control. The OMC supports network operators to manage the BSS and the NSS equipment. It includes fault, performance, and configuration management.

3.4.1 Mobile station

A mobile station (MS) consists of the mobile equipment (ME) and the subscriber identity module (SIM) as shown in Figure 3-2. The mobile equipment is a complex hardware device. The functionality includes radio transmitter/receiver, gaussian minimum shift keying (GMSK) modulation and demodulation, coding/decoding, and DTMF generation.

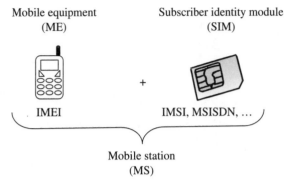

Figure 3-2 Mobile station components.

The firmware includes control logic, protocol stack for Call processing/ control, and mobility management. The international mobile equipment number (IMEI) uniquely identifies mobile equipment. The IMEI is burned within the module.

In today's environment, many of the commercially available MEs are capable of supporting multiple bands, i.e., 900, 1800, and 1900 MHz. This means these devices can be used almost universally.

The ME device, as such, does not have any subscriber information. As a stand-alone device, it cannot be used to make a call, with the exception of emergency calls. The subscriber identity module (SIM) is a smart card. The wireless service provider programs it with the subscription data. The SIM is used to store data related to PLMN (e.g., HPLMN country and network code), subscription (e.g., IMSI, MSISDN), roaming (forbidden networks, etc.), and security (PIN, PUK, etc). In addition, it can also store subscriber data such as SMS and phone numbers.

Mobile station identities

International mobile station equipment identity (IMEI). Every mobile equipment has a unique identifier, i.e., an international mobile station equipment identity (IMEI). The IMEI identifies the mobile equipment and not the subscriber. The IMEI is embedded within the hardware and cannot be changed.

The purpose of IMEI is to protect the mobile equipment from stealth. The wireless service providers maintain a list of all stolen MEs in a database (i.e., equipment identity register) and may deny services to such mobile equipment if they wish to.

As shown in Figure 3-3, the IMEI consists of:

Type approval code (TAC). This code is 6 digits/24 bits long. The TAC is issued by an authorized agency on successful testing for type approval.

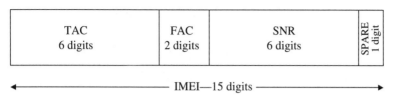

◄──────────────── IMEI—15 digits ────────────────►

Figure 3-3   IMEI format.

Final assembly code (FAC).   This code is 2 digits/8 bits long. This uniquely identifies the manufacturer of the mobile equipment.

Serial number (SNR).   Each mobile equipment is identified with a unique serial number within a TAC and FAC. The SNR is 6 digits/24 bits long.

The remaining 1 digit/4 bits are not currently used and are a "spare."

Mobile station ISDN number (MSISDN).   Each subscriber in a network is identified with a unique international number, i.e., a mobile station ISDN number. The wireless service provider assigns this number at the time of subscription. The format for the MSISDN is defined in the ITU-T E.164 recommendation. As shown in Figure 3-4, MSISDN is of variable length but limited to a maximum of 15 digits excluding prefixes.

Country code (CC).   Country codes are defined by the ITU-T. They can be 1 to 3 digits long. For example, the country code for the United States is 1, Japan is 81, and Ecuador is 593.

National destination code (NDC).   The country's telecommunication regulatory authority assigns an NDC to each PLMN. One PLMN may have more than one NDC assigned to it. This field may be 2 to 3 digits.

Subscriber number (SN).   The SN is a variable-length field.

The MSISDN number with national or international prefixes is used to call a subscriber. Figure 3-5 illustrates the dialing sequence if a subscriber in India calls a mobile subscriber registered in the Celcom network in Malaysia.

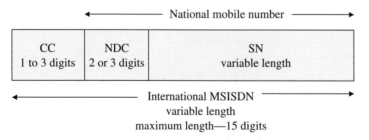

Figure 3-4   MSISDN format.

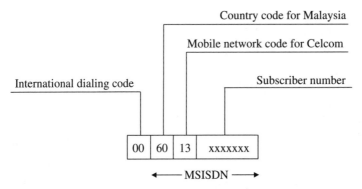

Figure 3-5   International dialing using MSISDN.

International mobile subscriber identity (IMSI).   International mobile subscriber identity (IMSI) is a unique identifier for a GSM subscriber in a PLMN. It is stored in the SIM and also in the HLR as part of the subscriber data. The HLR transfers IMSI information to the serving VLR on registration for temporary storage. ITU-T Recommendation E.212 defines the structure of the IMSI.

As shown in Figure 3-6, IMSI is 15 digits long and consists of MCC, MNC, and MSIN.

Mobile country code (MCC).   ITU-T E.212, Annexure-A, lists all the countries and assigned codes. The MCC is 3 digits long. For example, the MCC for Australia is 505, Germany is 262, and the United States is 310.

Mobile network code (MNC).   Each PLMN in a country is assigned a unique network code by a regulatory authority in the country. The MNC is 2 digits long. For example, in Singapore the assigned code for Singtel is 01, M1 is 03, and Starhub is 05.

Mobile subscriber identification number (MSIN).   The MSIN is a unique number within a PLMN to identify the subscriber.

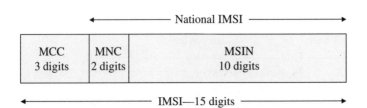

Figure 3-6   E.212 IMSI format.

Mobile station—temporary identities. In addition to MS identities described in the section, "Mobile station identities," the network assigns temporary identities to the MS during call processing to ensure security and facilitate roaming.

Temporary mobile subscriber identity. The temporary mobile subscriber identity (TMSI) is used in place of IMSI on the air interface to prevent a possible intruder from tracking a mobile subscriber. The IMSI is used only in cases where TMSI is not assigned, e.g., for initial registration while roaming in the other PLMN. The network assigns TMSI to all active subscribers. The TMSI is stored in the VLR along with the IMSI. The network may reallocate the TMSI during location update and call processing. On the network side, the VLR is responsible for TMSI management, i.e., assignment, storage, updating, and mapping with other identities. On the MS side, the TMSI is stored in SIM card. TMSI has only local significance within the serving MSC/VLR area. The network providers themselves decide the format of TMSI. However, the recommended length is four octets or less.

Mobile station roaming number. The mobile station roaming number (MSRN) is used during the mobile terminating call setup. This is a temporary identifier assigned by the VLR to a MS roaming in its serving area to facilitate call routing. The VLR manages the assignment and the reallocation of MSRN. As described in Section 3.6, under "Mobile Terminated Call," the VLR assigns the MSRN when a send routing information request is received from the HLR. The HLR then returns the MSRN to the GMSC, which uses it to route the call. This MSRN assignment is valid only during call setup and is released as soon as the call is established.

Figure 3-7 shows the format of the MSRN as defined by the GSM Recommendation. It consists of three parts:

Country code (CC). Country codes are defined by the ITU-T. It can be 1 to 3 digits. For example, the country code for the United States is 1, Japan is 81, and Ecuador is 593.

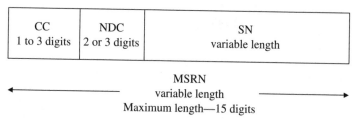

Figure 3-7 MSRN format.

National destination code (NDC). The country's telecommunication regulatory authority assigns the NDC to each PLMN. One PLMN may have more than one NDC assigned to it. This field may be 2 to 3 digits.

Subscriber number (SN). The SN is a variable-length field. In this case, this is a temporary subscriber number assigned by the serving MSC/VLR to an MS roaming in its coverage area.

3.4.2 Base station subsystem

As shown in Figure 3-1, the Base Station Subsystem (BSS) provides a connection between the MS and the GSM network via the air interface. It consists of two main functional entities.

- Base transceiver station (BTS)
- Base station controller (BSC)

Base station transceiver. The BTS contains the radio transceivers and antennas that provide radio interface to and from the Mobile Station. It defines the cell and handles radio link level protocols with the MS. The BTS can be considered as a counterpart of the MS within a GSM network. Each BTS may have up to 16 transceivers, each of which is allocated a different RF channel. The number of transceivers depends on the traffic-handling requirements in a particular cell. A number of BTSs are deployed in a network to achieve desired coverage. A BTS is usually installed at the center of a cell. The size of a cell is determined by the transmitting power of a BTS. The main tasks of a BTS include the following:

- Channel coding and decoding: speech, data, signaling
- Speech coding: half and full rate
- Ciphering
- GMSK modulation and demodulation
- Frequency hopping
- Time and frequency synchronization
- Timing advance and power control
- Radio link measurements and management
- Operation and maintenance

Base station controller. As the name implies, the BSC monitors and controls one or more base stations. The number of BTSs controlled by a BSC varies, depending on the technology adopted by a particular manufacturer. It provides a number of functions related to radio resources management (RRM) and mobility management (MM). The BSC connects with the BTS using the Abis interface, a description of which is given in the next section. The main tasks of a BSC include:

- Frequency administration
- Time and frequency synchronization
- Time delay measurements
- Handovers
- Power management
- Operation and maintenance

The same vendor usually supplies the BTS and BSC. This is because the BTS implementation is vendor specific. No standards are available for the internal design of a BTS.

Identities associated with the BSS

Location area identity. The PLMN is divided into one or more location areas (LAs). Each location area consists of one or more BTSs. The purpose of defining an LA is to optimize paging efficiency and location area updates. A mobile station moving from one cell to another in the same LA does not need to update its location. Also, if the network needs to contact the MS, it transmits only the paging message belonging to cells of the last known LA.

Each location area is identified by a unique location area identity (LAI). The format of an LAI is defined by the GSM Recommendations. Figure 3-8 shows the format and structure of an LAI.

Mobile country code (MCC). ITU-T E.212, Annexure A, lists all the countries and assigned codes. The MCC is 3 digits long. For example, the MCC for Australia is 505, Germany is 262, and the United States is 310.

Mobile network code (MNC). Each PLMN in a country is assigned a unique network code by the country's regulatory authority. The MNC

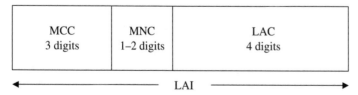

Figure 3-8 Location area identity.

is 2 digits long. For example, in Singapore the assigned code for Singtel is 01, M1 is 03, and Starhub is 05.

Location area code (LAC).   LAC identifies a location area within a GSM PLMN. It is 4 digits/16 bits long.

Cell global identity.   Each cell within a PLMN is assigned a 4-digit/16 bit code, which is known as the cell identity (CI). To make it distinct and unique on global basis, GSM Recommendations define cell global identity (CGI). The CGI (Figure 3-9) is a unique identifier of a cell within a location area. It is a combination of LAI and cell identity.

3.4.3   Network switching system

The role of the network switching system (NSS) is to set up call connections in a mobile environment. The NSS achieves this by using the following switching and database nodes:

- Mobile switching center (MSC and gateway MSC)
- Home location register (HLR)
- Visitor location register (VLR)
- Equipment identity register (EIR)
- Authentication center (AuC)

In addition, short message service center (SMSC) is required to support short message service.

In almost all the implementations, the HLR and the AuC functionality is implemented in one physical node usually referred to as the HLR/AuC. In the following sections, the HLR functionality as described includes AuC features.

Gateway mobile switching center.   The gateway mobile switching system (MSC) is the central component of an NSS. It is like a switching node/central office of the PSTN with additional capabilities to support mobility.

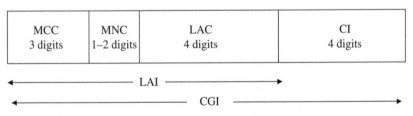

Figure 3-9   Cell global identity.

The GMSC supports all the GSM-specific interfaces defined in the next section. In addition, it also provides an interface to external networks, i.e., PSTN, ISDN, and other PLMNs. The services provided by GMSC include call setup, call routing, registration, authentication, location updating, handovers, and billing. The GMSC offers these services in conjunction with other NSS entities such as HLR, VLR, MSC, AuC, and EIR.

The main functions of a GMSC are:

- Registration, location updating
- Authentication and other security functions
- Paging
- Handover management
- Switching and signaling
- Billing
- BSS management

The number of GMSCs in a GSM network varies, depending upon the size of the network and its interfacing requirements with PSTN. Usually, one GMSC suffices. Each GMSC is associated with one or more HLR.

Mobile switching center. The MSC has the same functionalities as GMSC, with a few exceptions. The MSC has no direct interface with PSTN/other PLMNs. Also, it has no associated HLR. The number of MSCs in a network depends on the size of the network. The MSC/GMSC and the BSSs are connected via a standard and well-defined interface. Therefore, a BSS and an NSS from two different vendors can coexist.

Home location register and authentication center. The home location register (HLR) stores the identity and subscriber data of all the users registered in a GSM network. The information stored in the HLR includes permanent data such as the IMSI, MSISDN, authentication keys, permitted supplementary services, and some temporary data. Examples of temporary data stored in the HLR are the current address of the serving MSC/VLR and the roaming number to which the calls must be forwarded. The temporary data is required to support mobility. Table 3-1 lists some of the important data stored in the HLR/AuC.

Most of the vendors have implemented the authentication center (AuC) in the same node as the HLR. The AuC calculates and provides the authentication triplets, i.e., Kc, RAND, and signed response.

TABLE 3-1   Important Data Stored in the HLR/AuC

Subscription data	
IMSI	International mobile subscriber identity (IMSI), also stored in the SIM
Ki	Authentication key, also stored in the SIM
Service restrictions	Operator-determined barring (ODB)
SS	List of permitted supplementary services
MSISDN	Mobile subscriber ISDN number; one subscription may be associated with multiple MSISDNs

Security data	
A3/A8	Authentication algorithm, also stored in the SIM
RAND	Randomly generated number; it is used by the SIM as an input to calculate SRES
SRES	Signed RESponse
Kc	Ciphering key

Subscriber location	
VLR address	The address of the VLR in which the mobile is currently located
MSC address	The address of a serving MSC
LMSI	Local mobile subscriber identity

These are used for authentication and encryption over the radio channel. The HLR connects to an AuC, in cases where it is implemented as a separate node.

It is desirable to keep access time to retrieve the data from the HLR to a minimum. The access time directly impacts the call completion time. Hence, the number of HLRs in a GSM network is determined by several factors such as the access time and the amount of data stored for the number of subscribers. The HLR can be implemented as a distributed database for security, reliability, and performance reasons but logically one HLR exists per PLMN.

Visitor location register (VLR).   The visitor location register (VLR), like an HLR, is a database. It contains selected administrative data for all the mobiles currently located in a serving MSC associated with the VLR. As can be seen in Table 3-2, the permanent data stored in the VLR is the same as the data stored in the HLR. However, there are a few additional parameters mainly of temporary data type—for example, TMSI and MSRN.

The main function of a VLR is to support GMSC/MSC during authentication and call establishment. To enable this functionality, the HLR updates the VLR with the relevant subscriber information on a need basis. A MSC is always associated with only one VLR. However, a VLR can serve several MSCs.

TABLE 3-2   Important Data Stored in the VLR

Subscription data	
IMSI	International mobile subscriber identity (IMSI), also stored in SIM
TMSI	Temporary mobile subscriber identity (TMSI)
SS	List of permitted supplementary services
MSISDN	Mobile subscriber ISDN number

Security data	
RAND	Randomly generated number; it is used by SIM as an input to calculate SRES
SRES	Signed RESponse
CKSN	Ciphering key sequence number
Kc	Ciphering key

Subscriber location	
HLR address	The address of the VLR in which the mobile is currently located
MSC address	The address of serving MSC
LAI	Location area identity
MSRN	Mobile subscriber roaming number
LMSI	Local mobile subscriber identity

Equipment identity register (EIR).   The equipment identity register is a database that contains the list of IMEIs for all mobile equipment. As shown in Figure 3-10, it maintains three lists:

White list.   Contains IMEIs of all valid mobile equipment.

Black list.   Contains IMEIs for stolen mobile equipment. It also contains the list of IMEIs for mobile equipment that need to be barred because of technical reasons.

White list.   Contains the list of all IMEIs that need to be traced.

Figure 3-10   Equipment identity register.

The implementation of an EIR is not critical from the services point of view. It is an option and left to the network operators. From a technical point of view, the EIR is interrogated at the time of location update or any time during call setup.

3.5   GSM Interfaces and Protocols

The GSM specifications define the interaction between system components through well-defined interfaces and protocols. Figure 3-11 shows the interfaces between the GSM functional entities. Table 3-3 lists the GSM interfaces.

Figure 3-12 shows the protocol architecture used for the exchange of signaling messages on each interface. The protocols are layered according to the OSI Reference Model. It consists of the Physical Layer, Data Link Layer, and Layer 3. This Layer 3 is not the same as defined in OSI Layer 3. In GSM, the Layer 3 functions include call, mobility, and radio resource management. In the OSI model, these functions are provided by the higher layers. GSM reuses a few established protocols such as CCS7 MTP, TCAP, SCCP, ISUP, and ISDN LAPD protocols. The MAP and BSSAP are new protocols to support GSM specific needs.

3.5.1   Air interface

The air interface between the MS and the BTS is called Um. The GSM air interface is based on time division multiple access (TDMA) with frequency division duplex (FDD). TDMA allows multiple users to share a common RF channel on a time-sharing basis, while FDD enables different frequencies to be used in uplink (MS to BTS) and downlink (BTS to MS) directions. Most of the implementations use a frequency band of 900 MHz. The other derivative of GSM is called Digital cellular system

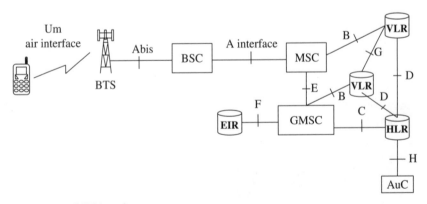

Figure 3-11   GSM interfaces.

TABLE 3-3   GSM Interfaces

Interface	Description
Um	MS↔BTS air interface
Abis	BTS↔BSC
A	BSC↔(G)MSC
B	(G)MSC↔VLR
C	(G)MSC↔HLR
D	VLR↔HLR
E	(G)MSC↔(G)MSC
F	(G)MSC↔EIR
G	VLR↔VLR
H	HLR↔AuC

1800 (DCM1800). It uses a frequency band of 1800 MHz. Table 3-4 lists the GSM frequency bands.

The used frequency band is divided into 200-kHz carriers or RF channels in both the uplink and downlink direction. Each RF channel is then further subdivided into eight different timeslots, i.e., 0 to 7, by TDMA techniques. A set of these eight timeslots is referred as a TDMA frame. Each frame lasts 4.615 ms. The physical channels are further mapped to various logical channels carrying user traffic and control information between the MS and the BTS. Table 3-5 describes the logical channels and their usages.

The following section describes the Um interface protocols used at the MS and the BTS side.

Physical layer. Layer 1, which is a radio interface, provides the functionality required to transfer the bit streams over the physical channels on the radio medium. The services provided by this layer to those above include:

- Channel mapping (logical to physical)
- Channel coding and ciphering
- Digital modulation
- Frequency hopping
- Timing advance and power control

Data link layer. Signaling Layer 2 is based on the LAPDm protocol, which is a variation of the ISDN LAP-D protocol. The main task of LAPDm is to provide a reliable signaling link between the network and the mobile station. The LAP-D protocol has been modified to adapt in the mobile environment. For example, LAPDm uses no flags for frame delimitation. The Physical Layer itself does the frame delimitation. This way, scarce radio resources are not spent on flag bits the bit.

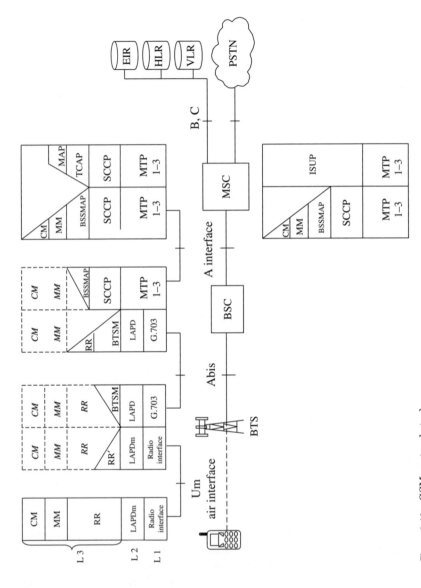

Figure 3-12 GSM protocol stack.

TABLE 3-4   GSM Frequency Bands

System	Direction	Frequency band (MHz)
GSM 900	Uplink	890–815
	Downlink	935–960
GSM/DCS1800	Uplink	1710–1785
	Downlink	1805–1880

Network layer. Signaling Layer 3 takes care of signaling procedures between an MS and the network. It consists of three sublayers with distinct signaling procedures.

- Radio resource management (RR)

- Mobility management (MM)

- Connection management (CM)

Radio resource management. Radio resource management (RR) comprises procedures required to establish, maintain, and release the dedicated radio connections. The RR sublayer functions include:

- Channel assignment and release

- Ciphering

- Modification of channel modes, e.g., voice and data

- Handover between cells

- Frequency redefinition to enable frequency hopping

- MS measurement reports

- Power control and timing advance

- Paging

- Radio channel access

The mobile station always initiates an RR session. For example, the RR procedures are invoked to establish an RR session in response to a paging message or to establish an outgoing call. As shown in Figure 3-13, the RR messages are transferred to BSC transparently, through the BTS. Table 3-6 lists RR messages.

Mobility management. The mobility management (MM) sublayer handles functions and procedures related to mobility of the mobile user. This includes procedures for:

- Authentication

- Location registration and periodic updating

TABLE 3-5   Logical Traffic and Control Channels

Traffic channels (TCH)	
TCH/F *Full-rate traffic channels* MS↔BTS	TCH/F carries subscriber information (speech/data) at a rate of 22.8 Kbps with a speech coding at around 13 Kbps.
TCH/H *Half-rate traffic channels* MS↔BTS	TCH/F carries subscriber information at a rate of 11.4 Kbps with a speech coding at around 7 Kbps.
Broadcast control channels (BCH)	
FCCH *Frequency correction channels* MS←BTS	This channel is broadcast by the BTS and carries information for the frequency correction of the MS. It is used in downlink direction only.
SCH *Synchronization channels* MS←BTS	This channel is broadcast by the BTS and carries information for frame synchronization of the MS. In addition it also carries the base station identity code (BSIC). It is used in downlink direction only.
BCCH *Broadcast control channels* MS←BTS	This channel carries broadcast information related to the BTS and the network. The information includes configuration details of common control channels (CCH) described below. It is used in downlink direction only.
Common control channels (CCH)	
PCH *Paging channel* MS←BTS	This is used to page an MS. It is used in downlink direction only.
RACH *Random access channel* MS→BTS	The MS uses this channel to request the allocation of a SDCCH. It is used in uplink direction only.
AGCH *Access grant channel* MS←BTS	The BTS allocates a SDCCH or TCH in response to the allocation request by the MS using this channel. It is used in downlink direction only.
Dedicated control channels (DCH)	
SDCCH *Stand-alone dedicated control channel* MS↔BTS	This channel is used for carrying signaling information between the BTS and a MS before allocation of a TCH. For example, SDCCH is used for carrying signaling messages related to update location and call establishment. This is a bidirectional channel.
SACCH *Slow associated control channel* MS↔BTS	This channel is always used in conjunction with a TCH or a SDCCH. The MS and the BTS use it to maintain an SDCCH or a TCH. In the uplink, the MS sends measurement reports to the BTS using this channel. In the downlink, the BTS transmits information to keep the mobile updated on recent changes in system configuration.
FACCH *Fast associated control channel* MS↔BTS	This channel is always associated to a TCH and is used to transfer signaling messages when a mobile is already involved in a call.

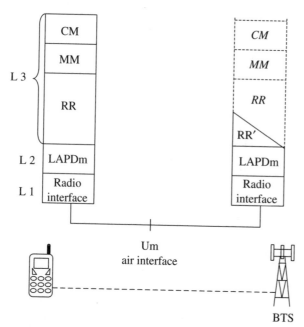

Figure 3-13   Air interface signaling protocols.

- Security
- TMSI reallocation
- IMSI detach/attach

As shown in Figure 3-13, the CM layer from the transmitting side uses the MM layer to establish RR connection and then transfers messages transparently across to the receiving side, that is MSC. Table 3-6 lists MM messages.

Connection management. The connection management (CM) sublayer contains the functions and procedures for call control. This includes procedures to establish, release, and access services and facilities. The CM consists of three sublayers, namely, call control (CC), supplementary services (SS), and short message services (SMS).

The call control sublayer provides procedures for ISDN call control. These procedures are based on ISDN call control procedures defined in the ITU-T Q.931 specification. However, the minor modifications are done to adopt these to mobile environment.

The supplementary service sublayer provides the procedures to support non-call-related supplementary services such as call forwarding and call waiting.

TABLE 3-6   Layer 3 Messages

RR messages	MM messages	CM messages
Channel establishment messages	*Registration messages*	*Call establishment messages*
ADDitional ASSignment	IMSI DETach INDication	ALERTing
IMMediate ASSignment	LOCation UPDating ACCept	CALL CONFirmed
IMMediate ASSignment EXTended	LOCation UPDating REJect	CALL PROCeeding
IMMediate ASSignment REJect	LOCation UPDating REQuest	CONnect
Paging messages	*Connection management messages*	CONnect ACKnowledge
PAGing REQuest Type 1	CM SERVice ACCept	SETUP
PAGing REQuest Type 2	CM SERVice REJect	EMERGency SETUP
PAGing REQuest Type 3	CM SERVice REQuest	PROGRESS
PAGing ReSPonse	CM SERVice ABOrt	*Call phase messages*
Handover messages	CM REeStablishment REQuest	MODify
ASSignment CoMmanD	*Security messages*	MODify REJect
ASSignment COMplete	AUTHentication REJect	MODify COMPlete
ASSignment FAILure	AUTHentication REQuest	USER INFOrmation
HANDover ACCess	AUTHentication ReSPonse	HOLD
HANDover CoManD		
HANDover COMplete	IDENTity REQuest	HOLD REJect
HANDover FAILure	IDENTity ReSPonse	HOLD ACKnowledge
PHYsical INFOrmation	TMSI REALlocation COMmand	RETRIEVE
Ciphering messages	TMSI REALlocation CoMPlete	RETRIEVEREJect
CIPHering MODe CoMmanD	*Other messages*	RETRIEVE ACKnowledge
CIPHering MODe COMplete	MM STATUS	*Call clearing messages*
Channel release messages	ABORT	DISConnect
CHANnel RELease		RELease
PARTial RELease		RELease COMplete
PARTial RELease COMplete		*Other messages*
System information messages		CONGESTion CONTROL
SYStem INFOrmation Type1		STATUS
SYStem INFOrmation Type2		STATUS ENQuiry
SYStem INFOrmation Type3		NOTIFY
SYStem INFOrmation Type4		START DTMF
SYStem INFOrmation Type5		STOP DTMF
SYStem INFOrmation Type6		START DTMF ACKnowledge
SYStem INFOrmation Type7		
SYStem INFOrmation Type8		
SYStem INFOrmation Type 2bis		STOP DTMF ACKnowledge
SYStem INFOrmation Type 5bis		START DTMF REJect

TABLE 3-6   Layer 3 Messages (*Continued*)

RR messages	MM messages	CM messages
Other messages CHANnel REQuest CHANnel MODe MODify CHANnel MODe MODify ACKnowledgment CLASSmark ENQuiry CLASSmark CHANGE FREQuency REDEFinition RR Status MEASurement REPort		

The short message service sublayer provides the procedures to support the short message transfer between the MS and the network. Table 3-6 lists CM messages.

3.5.2   Abis interface

Abis is the interface between the BSC and the BTS. Figure 3-14 illustrates the protocol stack on the Abis interface. The following sections describe each layer in detail.

Physical layer.   The physical layer, i.e., Layer 1, consists of a 2-Mb/s PCM30 link. This is based on ITU-T G.703 specifications. A PCM30 link consists of 32 multiplexed 64-Kb/s timeslots. Thirty timeslots carry speech or user data, and the remaining two timeslots are used for synchronization and signaling purposes. Most of the vendors support further

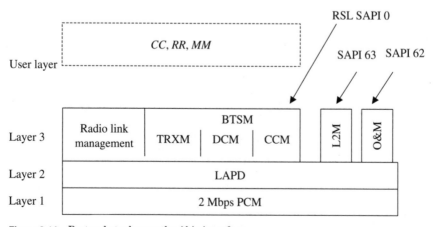

Figure 3-14   Protocol stack over the Abis interface.

division of each 64b/s timeslot into four 16-Kb/s timeslots. This has the obvious advantage of better link utilization. It also enables mapping of traffic channels at the Um interface directly to Abis. The traffic channels at the Um interface have almost the same data rates. A transcoder rate adoption unit (TRAU) is required to convert 64-Kb/s speech into 13-Kb/s GSM speech. The TRAU can be located at the BTS or BSC or MSC side. Generally, one 2-Mb/s link covers more than one BTS. The exact configuration of Abis links depends on the traffic requirements, TRAU location, and equipment capabilities.

Layer 2 protocol. The Data Link Layer, i.e., Layer 2, is based on the ISDN link access procedure on the D-Channel (LAP-D) protocol, with a few changes. The LAPD protocol is defined in ITU-T Q.920 and Q.921 specifications. The ITU-T Q.920 standard defines the general parameters of ISDN Layer 2. The ITU-T Q.921 defines the specifics of Layer 2. The main task of Layer 2 is to control the logical signaling links between a BSC and its connected BTSs. It also ensures error-free transmission of information between communicating entities.

Each BTS is connected on the Physical Link with the controlling BSC. However, the BTS have several logical LAPD data links over a physical link. The logical links are provided for Layer 3 information transfer and O&M of the BTS equipment and the links themselves. Each logical link is uniquely identified with a service access identifier (SAPI) and terminal equipment identifier (TEI) combination. The SAPI and TEI are parts of the address field in a LAP-D frame. The LAP-D frame is shown in Figure 3-16 and will be explained later in this section.

The SAPI identifies the Layer 3 protocol. The SAPI is 6 bits long and can have a value from 0 to 63. However, in GSM only three values, as given in Table 3-7, are used.

The TEI identifies one transceiver (TRX). The TEI is 7 bits long and hence can have a value from 0 to 127. GSM uses TEI values 0 to 63 for fixed TRX addresses. The values from 64 to 126 are used for additional TRX addresses in cases where TRX needs more than one signaling link.

TABLE 3-7   SAPI Values Used in GSM

SAPI (decimal)	Description
0 Radio signaling link (RSL)	This link is used to transfer Abis Layer 3 messages between BTSs and BSC. In addition, it serves the traffic management procedures of Layer 2.
62 Operation & maintenance link (OML)	This link is used to transfer BTS O&M messages.
63 Layer 2 management link (L2ML)	It is used for management of logical data link sharing of a physical connection.

Figure 3-15 shows the concept of uniquely identifying a logical data link using SAPI and TEI. One signaling link between the BTS and the BSC consists of three logical channels, RSL, OML, and L2ML, each of which is uniquely identified with a combination of SAPI and TEI.

LAP-D frame structure. Figure 3-16 shows a LAP-D frame structure. The flags indicate the beginning and the end of a frame. For consecutive frames, one frame is used to indicate the end of a first frame and the beginning of the next frame. A flag is 0111 1110 (hex 7E). In order to avoid repetition of this pattern within the information field, a zero is inserted after every five consecutive ones. This is called bit stuffing.

The two-octet address field, also known as the data link control identifier (DLCI), includes SAPI and TEI. The function of SAPI and TEI, as described in the previous section, is to identify logical data links. Each octet in the address field has one address extension (EA) bit. In the first octet, it is set to zero, indicating that one more address octet is to follow. The EA bit of the second octet is set to 1, indicating that it is the last octet of the address field. The command/response (C/R) bit is used to differentiate the commands from the responses. The BTS (user side) sets the C/R bit to one for responses and resets it to zero for commands. The BSC (network side) does the opposite, i.e., it resets the C/R bit when it sends a response and sets it when it sends a command.

There are three different formats of the control field.

Information transfer format (I frames). I frames control the transfer of the LAPD frame's information field to Layer 3. I frames use N(S)

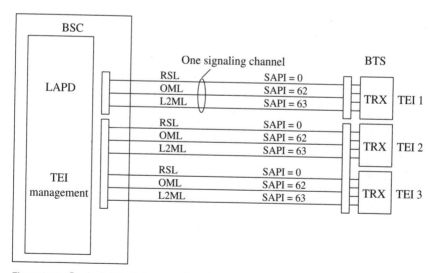

Figure 3-15 Logical data links over the Abis interface.

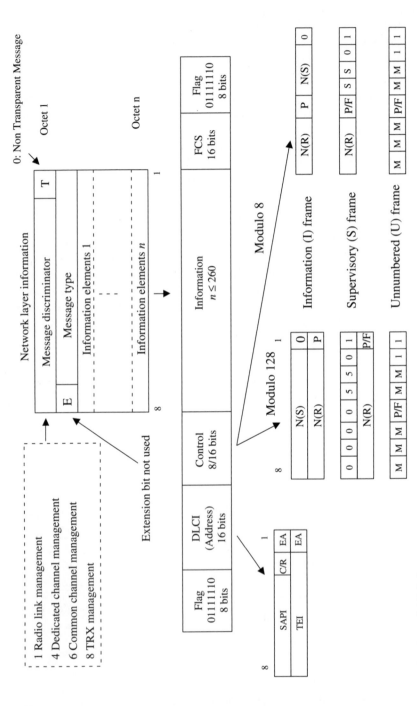

Figure 3-16 LAP-D frame structure.

(send sequence number), N(R) (receive sequence number), and P/F (poll/final bits) to number the frames and acknowledge correct receipt of frames.

Supervisory format (S frames).   S frames handle Layer 2 flow control management, such as acknowledging the I frames and requesting retransmission and temporary suspension of I frames. N(R) and P/F bits are used in these frames.

Unnumbered information and control format (U frames).   U frames provide additional transfer capabilities during unacknowledged transfer service or additional unacknowledged transfer service. N(S) and N(R) bits are not used. Only P/F bits are used.

Figure 3-16 shows the two different formats of the control field. The length (8 or 16 bits) of control field depends on the frame type and also on the sequence numbering used, i.e., modulo 8 or modulo 128.

Table 3-8 describes the different format and frame types. Table 3-9 describes the functions of different frame types.

The information filed is of variable length and carries Layer 3 information. A maximum of 260 octets can be sent over LAPD. The information field is present in all I frames and U frames that transfer information, i.e., UI frames. It is not present in S and U frames with only one exception, i.e., FRMR.

The frame check sequence (FCS) is used to detect errors in a frame. It is a 16-bit cyclical redundancy checksum (CRC) defined by ITU-T.

Layer 3 protocol.   At Layer 3, between BSC and BTS, two different types of message flow take place, i.e., transparent and nontransparent messages.

TABLE 3-8   LAP-D Frame Formats

Control field format	Name of frame	Type of frame	Control field length (octets)
I frame	Information (I)	Command	2
S frames	Receiver ready (RR)	Command response	2
	Receiver not ready (RNR)	Command response	2
	Reject (REJ)	Command response	2
U frames	Set asynchronous balanced mode extended (SABME)	Command	1
	Disconnect mode (DM)	Response	1
	Unnumbered information (UI)	Command	1
	Disconnect (DISC)	Command	1
	Unnumbered acknowledgement (UA)	Response	1
	Frame reject (FRMR)	Response	1
	Exchange identification (XID)	Command	1

TABLE 3-9   LAP-D Frame Functions

Frame	Functions
Information (I)	I frame carries Layer 3 information across a data link connection during acknowledged transfer service.
Receiver ready (RR)	RR frames are used to indicate: ■ Layer 2 entity is ready to receive ■ Acknowledgment of previously received I frame
Receiver not ready (RNR)	It is used to indicate that a data link layer entity is busy and no more I frames can be accepted.
Reject (REJ)	A reject command frame requests retransmission of I frames starting with a frame numbered N(R). As a response, an REJ frame indicates the clearance of a busy condition.
Set asynchronous balanced mode extended (SABME)	SABME frame begins a data link connection for acknowledged information transfer service.
Disconnect mode (DM)	The transmitting side uses the DM frame to indicate that it can no longer maintain the Layer 2 connection.
Unnumbered information (UI)	UI frame carries Layer 3 information across a data link connection during unacknowledged transfer service.
Disconnect (DISC)	The transmitting side indicates its intention to tear down the Layer 2 connection by sending a DISC frame.
Unnumbered acknowledgement (UA)	A UA frame is used as a response to a SABME or DISC frame.
Frame reject (FRMR)	Unlike a reject frame, the FRMR is used to report an error condition that cannot be recovered by retransmission of frame. For example, a protocol error detected in a Layer 3 message cannot be set right simply by retransmission of a frame.
Exchange identification (XID)	Not used in GSM.

Transparent messages pass through the BTS without any decoding and action. These messages are from the MS and intended to be for the BSC/MSC or the other way around. The CM and the MM messages are examples of the transparent messages. The BTS does not process these messages. However, the RR layer contains messages of both types. The nontransparent messages, in this case, are those related to radio equipment and need to be handled by the BTS. The BTS management layer at the BTS interprets these messages and performs actions. An example of nontransparent RR messages is the ciphering message, where the ciphering key is sent to the BTS only, not to the MS.

As shown in Figure 3-15, the signaling channel between the BTS and the BSC carries three logical channels: RSL, OML, and L2ML. Table 3-7 lists the SAPI assigned to each logical channel. The RSL, which is assigned SAPI0, carries user signaling, i.e., all messages related to connection setup, release, SMS, and supplementary services (SS). The messages sent over RSL are divided into four groups.

Radio link layer management (RLM). The RLM contains the messages related to status and control of Layer 2 connection between the BTS and the BSC.

Common channel management (CCM). The CCM contains the messages that carry common control channel (CCCH) signaling data to and from the air interface.

TRX management (TRXM). The TRXM contains the messages that are related to TRX management.

Dedicated channel management (DCM). The DCM contains messages related to status and control of Layer 1 of the air interface.

Figure 3-17 shows the Layer 3 message structure for a transparent message. In this example, the LAP-D frame is carrying a Layer 3 CM service request transparently. The message discriminator is used to distinguish between RRM, TRXM, CCM, and DC messages. In the example shown, the message discriminator is set to 1, indicating an RRM message. The bit T is set to 1, indicating a transparent message. The protocol discriminator is used to discriminate between RR, MM, and CM (CC and SMS).

Table 3-10 lists RLM, CCM, TRXM, and DCM messages. The uppercase letters in the message name are mnemonics used in context and protocol presentations.

3.5.3 A interface

The A interface is the interface between the BSC and the MSC. At the physical layer, it uses a 2-Mbps PCM30 link. One or more 64-Kbps timeslots are used to carry signaling information. Typically, more than one 2-Mbps link is required to handle the traffic between the BSC and the MSC.

The Base Station System Application Part (BSSAP) is a GSM-specific protocol designed for signaling over the A interface. The BSSAP uses CCS7 MTP and SCCP transport and addressing services for the signaling message transfer. The BSSAP supports both connectionless and connection-oriented services provided by the SCCP. The connectionless services are used to support global procedure such as PAGING (for MS) and RESET (a circuit). The connection-oriented services are used for dedicated procedures such as handover and assignment procedures. The BSSAP supports messages sent between the MSC and the BSS, as well as transparent message transfer between the MSC and the MS. To enable this functionality, the BSSAP is divided into two parts, i.e., the Base Station Subsystem Management Application Part (BSSMAP) and Direct Transfer Application Part (DTAP).

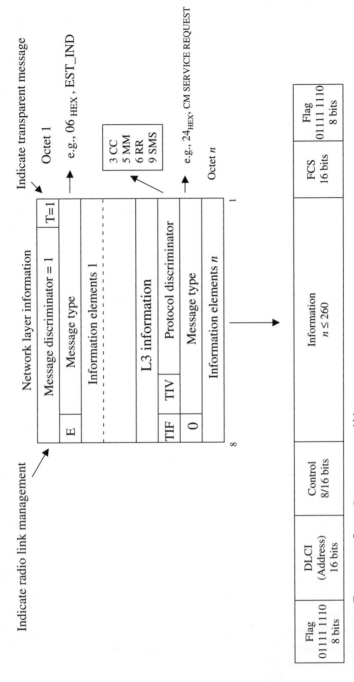

Figure 3-17 Transparent Layer 3 message over Abis.

TABLE 3-10   Abis Layer 3 Messages

RLM	CCM	TRXM	DCM
DATA REQuest '	BCCH INFOrmation	RF REsource INDication	CHANnel ACTivation
DATA INDication	CCCH LOAD INDication	SACCH FILling	CHANnel ACTivation ACKnowledge
ERROR INDication	CHANnel REQuired	OVERLOAD	CHANnel ACTivation Negative ACKnowledge
ESTablish REQuest	DELETE INDication	ERROR REPORT	CONNection FAILure INDication
ESTablish CONFirm	PAGING CoMmand		DEACTivate SACCH
ESTablish INDication	IMMEDIATE ASSign CoMmand		ENCRyption CoMmand
RELease REQuest	SMS Broadcast REQuest		HANDOver DETection
RELease CONFirm			MEASurement RESult
RELease INDication			MODE MODIFY REQuest
UNIT DATA REQuest			MODE MODIFY ACKnowledge
UNIT DATA INDication			PHYsical CONTEXT REQuest
			PHYsical CONTEXT CONFirm
			RF CHANnel RELease
			MS POWER CONTROL
			BS POWER CONTROL
			PREPROCess CONFIGure
			PREPROCessed MEASurement RESult
			RF CHANnel RELease ACKnowledge

Figure 3-18 illustrates the protocol stack on the MSC and the BSC side. The MM and CM sublayer signaling information from the MS is routed to the BSS transparently over the signaling channels (FACCH, DACCH, SACCH). From the BSS, this information is relayed to the MSC by DTAP. The DTAP uses SCCP logical connection to transfer the information to the peer MM or CC entity in the MSC.

BSS management application part. The BSSMAP data are part of Layer 3 and carry messages related to radio resource and the BSC management. The BSSMAP process within the BSC controls the radio resources in response to the instructions given by the MSC. Examples of BSSMAP

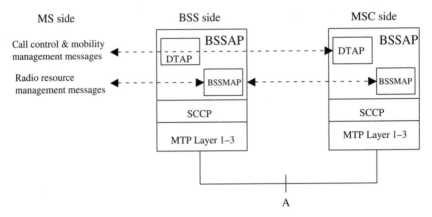

Figure 3-18   A-interface signaling protocol.

messages are paging, handover request, reset, and block. Table 3-11 lists the BSSMAP messages.

Direct transfer application part.   The DTAP data is user information and carries messages related to call control and mobility management between two users, i.e., MS and any subsystem of NSS such as the MSC. The messages are transparent to BSS except for a few exceptions. These exceptions are location update request, CM service request, and IMSI detach indication. These messages are partially processed by the BSC to add necessary information required for other entities to process these requests.

The DTAP messages are identical to the transparent MM and CM messages listed in the previous section.

3.5.4   Inter-MSC signaling

For call control, an MSC may have signaling interfaces to other MSC, GMSC, PLMN, and PSTN. Figure 3-19 shows the protocol stack used on these interfaces. ITU-T ISUP or TUP protocol is used for call setup and supervision. TUP is an old protocol and may not be used in newer implementations. ISUP is used for both speech and data call setup. ISUP relies on the MTP protocol for transportation, addressing, and routing of call control messages. ISUP and MTP protocols are described in Chapter 2.

The MSC also has interfaces with the VLR, HLR, EIR ,GMSC, and interworking MSCs for non-circuit-related call control. Figure 3-20 shows this interface and protocol stack. The GSM Mobile Application Part (MAP) has been specifically designed for transfer of non-circuit-related signaling information between MSCs and between MSCs and databases. MAP relies on TCAP capabilities to establish non-circuit-related communication between two entities in the signaling network to exchange data and

TABLE 3-11   BSSMAP Messages

Message type	ID (hex)	Direction	Remarks
		Assignment messages	
Assignment request	01	MSC→BSC	ASS REQ is sent from the BSC to the MSC to request a traffic channel on A and air interface.
Assignment complete	02	BSC→MSC	ASS COM is a positive response from the BSC.
Assignment failure	03	BSC→MSC	ASS FAIL is a negative response from the BSC.
		Handover messages	
Handover request	10	MSC→BSC	The MSC sends HND REQ to the new BSC that is the target for handover.
Handover required	11	BSC→MSC	The BSC sends HND RQD to the MSC in case of inter-BSC or inter-MSC handover.
Handover request acknowledge	12	BSC→MSC	HND REQ ACK is the acknowledge for the previously received HND REQ.
Handover command	13	MSC→BSC	HND CMD provides the information to the BSC related to the new radio channel resource to which the MS should switch.
Handover complete	14	BSC→MSC	HND CMP indicates successful handover
Handover failure	16	BSC→MSC	HND FAIL indicates unsuccessful handover.
Handover performed	17	BSC→MSC	HND PERF indicates that the BSC has performed the intra-BSC handover.
Handover candidate enquiry	18	MSC→BSC	HND CND ENQ is sent by the MSC to get the list of MSs in a particular cell that could be handed over to another cell to reduce the load on a target cell.
Handover candidate response	19	BSC→MSC	HND CND RES is a response to the HND CND ENQ message.
Handover required reject	1A	MSC→BSC	HND RQD REJ indicates an unsuccessful response to HND_RQD message.
Handover detect	1B	BSC→MSC	The BSC sends HND DET to the MSC updating changes in the radio resource on receiving the HND DET message from the BTS.
		Release messages	
Clear command	20	MSC→BSC	CLR CMD is used to release a traffic channel allocated to a specific MS.

(Continued)

TABLE 3-11 BSSMAP Messages *(Continued)*

Message type	ID (hex)	Direction	Remarks
Release messages			
Clear complete	21	BSC→MSC	CLR CMP is the confirmation of resource release in response to CLR CMD message.
Clear request	22	BSC→MSC	The BSC sends CLR REQ to the MSC on detecting any severe problem with an existing connection to an MS.
SAPI "n" reject	25	BSC→MSC	The BSC sends SAPI REJ message to the MSC on receiving a message with SAPI not equal to zero but for which no Layer 2 connection exists.
Confusion	26	BSC↔MSC	This message is sent to in response of a message which can not be treated correctly by the receiving entity and for which another failure message can not substitute.
General messages			
Reset	30	BSC↔MSC	Send by the MSC to the BSC (or vice versa) if the sending entity detects any fatal error in communication data.
Reset acknowledge	31	BSC↔MSC	RES ACK is sent to confirm that the RESET message was received.
Overload	32	BSC↔MSC	The BSC sends overload message to indicate the overload situation in a BTS or whole BSS. The MSC sends overload message to the BSC to indicate processor overload within the MSC.
Reset circuit	34	BSC↔MSC	RES CIRC is used to initialize a single circuit between the BSC and the MSC.
Reset circuit acknowledge	35	BSC↔MSC	Response to RES CIRC on a successful reset of a circuit.
MSC invoke trace	36	MSC→BSC	Request to start a trace of a single connection. MSC INV TRC will enable tracing of messages in the direction from MSC to BSC.
BSS invoke trace	37	BSC→MSC	Request to start a trace of a single connection. BSS INV TRC will enable tracing of messages in the direction from BSC to MSC.
Terrestrial resource messages			
Block	40	BSC→MSC	The BSC requests the MSC to block a single channel, using BLO message.
Blocking acknowledge	41	MSC→BSC	The MSC acknowledges the blocking of a channel by sending BLO ACK message to the BSC.
Unblock	42	BSC→MSC	UNBLO message is used to cancel the blocking.

TABLE 3-11   BSSMAP Messages *(Continued)*

Message type	ID (hex)	Direction	Remarks
		Terrestrial resource messages	
Unblocking acknowledge	43	MSC→BSC	Acknowledgment of BLO message.
Circuit group block	44	BSC→MSC	The BSC requests the MSC to block the multiple channels or complete PCM link, using CIRC GRP BLO message.
Circuit group blocking acknowledge	45	MSC→BSC	The MSC acknowledges the blocking of the multiple channels/complete link by sending CIRC GRP BLO ACK message to the BSC.
Circuit group unblock	46	BSC→MSC	CIRC GRP UNBLO message is used to cancel the blocking of multiple channels or complete PCM link.
Circuit group unblocking acknowledge	47	MSC→BSC	Acknowledgment of CIRC GRP UNBLO message.
Unequipped circuit	48	BSC↔MSC	This message is sent by an entity to its peer entity that it is using one or more circuit identity code which are unknown.
		Radio resource messages	
Resource request	50	MSC→BSC	The MSC requests the BSC to provide information on available radio resources.
Resource indication	51	BSC→MSC	Response to RES REQ.
Paging	52	MSC→BSC	The MSC sends this message to locate the MS in case of MTC.
Cipher mode command	53	MSC→BSC	The MSC sends a CIPHER MOD CMD to the BSC to instruct to start ciphering on air interface.
Classmark update	54	BSC→MSC	The MS sends CLS MRK UPD to the MSC via BSC to update the classmark changes, if any.
Cipher mode complete	55	BSC→MSC	The BSC sends this message to the MSC in response to the previously received CIPH MOD CMD to indicate that ciphering has successfully initiated on air interface.
Queuing indication	56	BSC→MSC	Indicates the delay in assignment of traffic channels because of no availability of resources in response to previously received ASS REQ or HND REQ.
Complete Layer 3 information	57	BSC→MSC	The CL3I message contains the initial message received from the MS for which SCCP connection at A interface can be set up. The typical example of such an initial message is CM SERV REQ.
Classmark request	58	MSC→BSC	This requests MS to send its classmark information.
Cipher mode reject	59	BSC→MSC	An unsuccessful response to the previously received CIPH MOD CMD.

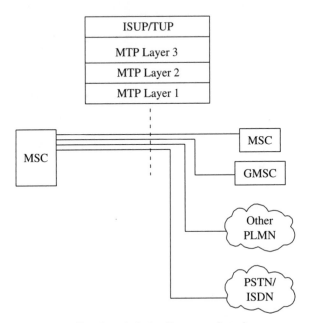

Figure 3-19   Circuit-switched call—protocol stack.

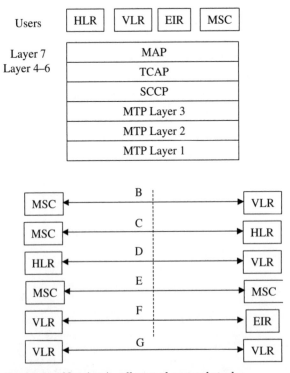

Figure 3-20   Noncircuit call control protocol stack.

control information. TCAP in turn relies on SCCP addressing and MTP transport capabilities. SCCP and TCAP protocols are described in Chapter 2.

MAP is also used between the VLR and the HLR. The B interface between the MSC and the VLR is an internal interface in most of the implementations.

The communication between applications and the MAP is done via MAP services or local operation codes defined in the GSM MAP specification. Table 3-12 lists local operation codes and their functions.

The local operation codes defined for the MAP operation on the B interface (MSC-VLR) are not listed in Table 3-12. The B interface is an internal interface and is not visible in most implementations. The GSM recommendations do not encourage the use of this interface and have not updated specifications for this interface since Release 99.

3.6 Scenarios

3.6.1 Mobility management

Location update. GSM allows a subscriber to move throughout the coverage area with a capability to make or receive calls. This is possible because the GSM network continuously tracks the movement of a mobile and updates its location all the time. There are three different location-updating scenarios.

- Attach/initial registration, used when a mobile is turned on

- Normal/forced registration, initiated when a mobile station moves to a new location area

- Periodic updates, regular location updates on expiry of a timer. This is required to track those mobile stations the locations of which may be lost because of some reason.

Figure 3-21 shows a location update (LU) procedure involving a mobile station, BTS, BSC, and MSC/VLR. In cases where the LU procedure is triggered because of VLR area change, the HLR and the old MSC/VLR are also involved. In GSM networks, where EIR is implemented and the IMEI check is turned on, additional signaling for verification is involved but is not shown in the figure.

The mobile station initiates the update location procedure by sending a **channel request** message on the random access channel toward the BTS. The reason for this request is also indicated in the request message. The reason, or establishment cause, in this case is update location. Other examples of establishment causes are voice call establishment and emergency call establishment. The BTS then sends a **channel required** message to the BSC.

The BSC checks the availability of a free SDCCH and activates it for the MS, using a **channel activation** message toward the BTS. This message

TABLE 3-12 MAP Local Operation Code

Operation	Op code	Description
		Mobility management
updateLocation	2	This operation code is send by the VLR to update location information stored in the HLR.
cancelLocation	3	The HLR uses this operation code to delete a subscriber record from a VLR/SGSN. For example, when a MS moves to a new VLR area, the HLR informs the old VLR to remove data for the MS.
sendIdentification	55	The new VLR uses this to retrieve the IMSI and authentication data from the old VLR for a subscriber registering in that VLR.
purgeMS	67	The VLR/SGSN sends this to the HLR to request the HLR to mark this MS as not reachable in case of MT call, MT-SMS, or network-initiated PDP context.
updateGprsLocation	23	The SGSN sends this message to update location information stored in the HLR.
noteMMEvent	89	The VLR/SGSN sends this message to the gsmSCF to indicate that a mobility management event has been processed successfully
prepareHandover	68	This is sent by an MSC that needs to hand over a call to another MSC.
sendEndSignal	29	This message is sent by an MSC to another MSC during the process of handover, indicating that radio path has been established to the MS and the recipient MSC can now release the resources.
processAceessSignaling	33	This message is sent by an MSC to pass information received on A interface or Iu interface to another MSC.
forwardAccessSignaling	34	This message is sent by an MSC to forward information to the A interface or Iu interface of another MSC.
prepareSubsequentHandover	69	This message is used by an MSC to inform another MSC that it has been decided to handover to a third MSC.
sendAuthenticateInfo	56	This message is used by the VLR/SGSN to request authentication information from the HLR.
authenticationFailureReport	15	This message is used between VLR/SGSN and the HLR to report authentication failure.
checkIMEI	43	This is used between MSC/SGSN and EIR to request IMEI check.
insertSubscriberData	7	This message is send by an HLR to update a VLR/SGSN with certain subscriber data.
deleteSubscriberData	8	This message is used by an HLR to remove certain subscriber data from the VLR/SGSN.

TABLE 3-12   MAP Local Operation Code *(Continued)*

Operation	Op code	Description
Mobility management		
reset	37	This operation code is used by the HLR in case of restart after failure. This is sent to all VLRs and SGSNs for which mobile stations were registered before restart.
forwardCheckSSIndication	38	This operation code is used by the HLR after restart to MS to indicate that a supplementary service parameter may have been altered. This message is sent by the HLR via VLR/MSC. This is an optional capability and implementation dependent.
restoreData	57	This is used by the VLR to synchronize the data with the HLR for a particular IMSI in certain cases. For example, this is invoked by the VLR if it receives a provideRoaming-Number operation code from the HLR for an IMSI that is unknown to the VLR.
anyTimeInterrogation	71	This is used by the gsmSCF to interrogate the HLR for the current state or location of an MS.
provideSubscriberInfo	70	This is used by the HLR to retrieve the MS state or location from the VLR/SGSN.
anyTimeModification	65	This is used by the gsmSCF to request subscription information from the HLR.
anyTimeSubscription-Interrogation	62	This is used by the gsmSCF to request subscription information from the HLR.
noteSubscriberDataModified	5	This is used by the HLR to inform the gsmSCF that the subscriber data has been modified.
Operation and maintenance		
activateTraceMode	50	When an operator executes a command in an OMC to trace a subscriber, the HLR requests that the VLR/SGSN activate subscriber tracing. The VLR/SGSN waits for the subscriber to become active before tracing can begin in the MSC/SGSN.
deactivateTraceMode	51	This is to deactivate the subscriber tracing in the MSC/SGSN. This operation code is sent by the HLR to the VLR/SGSN on request from the OMC.
sendIMSI	58	On request from the OMC, the VLR in a VPLMN requests the HPLMN HLR to get the IMSI for the subscriber whose MSISDN is known.
Call handling		
sendRoutingInformation	22	The GMSC/gsmSCF uses this operation in case of a mobile terminating call to interrogate the HLR for routing information. The HLR returns information such as VMSC address and the MSRN assigned to the subscriber.

(Continued)

TABLE 3-12   MAP Local Operation Code (*Continued*)

Operation	Op code	Description
	Call handling	
provideRoamingNumber	4	On receiving sendRoutingInfo request from the GMSC/gsmSCF, the HLR requests the VLR where the subscriber is currently registered to send assigned MSRN.
	Supplementary services	
registerSS	10	The VLR uses this operation code to enter the supplementary service data for a specific subscriber in the HLR. For example, if a subscriber registers any SS service on the phone, the registration will be passed to the HLR by the MSC and VLR transparently. The SS code parameter in this operation determines the supplementary service to be registered.
eraseSS	11	This is used to delete SS-related data for a specific subscriber in the HLR.
activateSS	12	This is used between VLR and HLR to activate an SS for a specific subscriber.
deactivateSS	13	This is used between VLR and HLR to deactivate an SS that was previously activated for a specific subscriber.
interrogateSS	14	This is used by the VLR to interrogate the status a specified supplementary service in the HLR.
registerPassword	17	This operation is invoked by the VLR toward the HLR on requests from the MS to register a new password or change an existing password.
getPassword	18	This operation is invoked by the HLR to get the password from the MS. This may be required if an MS tries to register an SS that requires a password from the subscriber. The VLR acts transparently and relays the message to the MSC.
processUnstructured-ssRequest	59	This operation is used to handle unstructured supplementary service between two entities. Unstructured SS is an additional means to build new supplementary services not defined by the GSM specifications. This request is sent between MSC and VLR, VLR and HLR, HLR and HLR, or HLR and gsmSCF.
unstructuredSSRequest	60	This operation is used by the requesting entity to get the information from the MS in connection with the handling of an unstructured supplementary service handling. This request is sent between MSC and VLR, VLR and HLR, or HLR and gsmSCF.
unstructuredSSNotify	61	This operation is used by the requesting entity to send a notification to the MS in connection with an unstructured supplementary service handling. This

TABLE 3-12   MAP Local Operation Code (Continued)

Operation	Op code	Description
Supplementary services		
ssInvocationNotify	72	request is sent between MSC and VLR, VLR and HLR, or HLR and gsmSCF. The MSC sends this operation code to the gsmSCF when a subscriber invokes one of the following supplementary services: ■ Call deflection (CD) ■ Explicit call transfer (ECT) ■ Multiparty (MP)
registerCcEntry	76	When a subscriber registers for call completion supplementary service, this operation code is send by the VLR to register the data in the HLR.
eraseCcEntry	77	This is used by the VLR to delete call completion supplementary service data for a specific subscriber in the HLR.
Short message service management		
sendRoutingInfoForSM	45	In case of MT-SMS, the GMSC sends this message to the HLR to get routing information needed for routing the short message to the serving MSC.
moForwardShortMessage	46	This is used by the serving SGSN/MSC to forward a mobile-originated short message to the interworking MSC for ultimate submission to the SMSC.
reportSmDeliveryStatus	47	This operation code is used between the GMSC and the HLR in case of unsuccessful delivery of SM to a MS. The HLR, on receiving this message, sets the message waiting flag for the MS. Once the serving VLR informs the HLR of contact renewal with the MS by sending the readyForSM message, the HLR resets the flag and sends an alertService Center message to the interworking MSC.
readyForSM	66	See description for reportSmDeliveryStatus.
alertServiceCenter	64	See description for reportSmDeliveryStatus.
informServiceCenter	63	This is used by the HLR to inform GMSC that the status of the subscriber for whom routing information is requested is not reachable.
mtForwardShortMessage	44	This is used by the GMSC to forward a mobile terminating short message to the serving MSC/SGSN.
Network requested PDP context activation		
sendRoutingInfoForGPRS	24	This operation is invoked by the GGSN to get the routing information from the HLR.
failureReport	25	This is used by the GGSN to inform the HLR that network-initiated PDP context activation was not successful.
noteMsPresentForGPRS	26	This is used by the HLR to inform the GPRS of the availability of an MS.

(Continued)

TABLE 3-12   MAP Local Operation Code (*Continued*)

Operation	Op code	Description
	Location service management	
sendRoutingInforforLCS	85	This operation code is sent to the HLR by the GMLC to retrieve the routing information needed for routing a location service request to serving MSC/SGSN.
provideSubscriberLocation	83	This is used by the GMLC to request the location of a specified MS from the SGSN/VLR.
subscriberLocationReport	86	This message is sent by the SGSN/MSC to the GMLC to provide the location of a specified MS. This is invoked in response to the message provideSubscriberLocation.

contains the information on reserved channel type and number. On receiving the acknowledgment from the BTS on channel activation, the BSC sends the **immediate assignment command** on the AGCH. The MS then moves to the assigned SDCCH and activates a Layer 2 connection by sending a LAPD **SABM** message, which also includes **location update request**. The location update request message contains several information elements, including the type of update location (attach, periodic, or normal), old LAI, IMSI, or TMSI. The BTS confirms the LAPD connection by sending an **unnumbered acknowledgment**. The BTS then passes the location update request to the BSC in an **establish indication** message.

The BSC processes the establish indication message, adds necessary information such as new LAI, and then establishes the SCCP connection (connection-oriented) with the MSC by sending a **connection request** (CR) message. The CR message also carries a location update request. The MSC acknowledges the CR message by sending a **connection confirmed** (CC) message.

The MSC/VLR sends the **authentication request** to the BSC, using the previously established SCCP connection. The BSC passes this request to the MS transparently. The authentication request message contains two important parameters: a 128-bit random number (RAND) and a 3-bits ciphering key sequence number (CKSN). The SIM within MS uses RAND and Ki, which is stored in SIM to calculate signed response (SRES) based on the A3 algorithm. The MS sends the SRES value as a parameter within the **authentication response** message to the MSC/VLR.

The VLR compares the SRES received from the MS to the SRES value available within it as a result of a previous send authentication procedure with the HLR/AuC. The MS is successfully authenticated if the two values match.

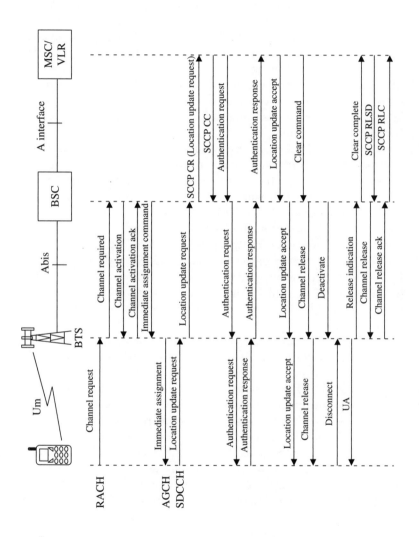

Figure 3-21 Location update procedure.

On successful authentication, if the ciphering is active, then the MSC/VLR initiates a ciphering mode setting procedure by sending a **ciphering mode command** to the BTS. This message contains the cipher key (Kc) and algorithm to calculate ciphering key. The BTS extracts and stores the cipher key before passing this message to the MS. The MS, on receiving this message, calculates the value of Kc, using Ki and RAND. The Ki and RAND are the same parameters, which were received previously during the authentication procedure. The algorithm used for the Kc computation is A8.

From this point onward, the MS starts ciphering all the data toward the BTS, using the A5 algorithm, as shown in Figure 3-22. The MS indicates to the BTS that ciphering has started by sending a **ciphering mode complete** message. The network also started ciphering all the data toward the MS. The BTS then sends ciphering mode complete in the Layer 3 **data indication** (DI) frame.

In cases where an IMEI check is enabled, then MSC/VLR requests the MS to provide its IMEI by sending an **identity request** message. The IMEI check, however, can be performed any time during the Update Location scenario. The MS responds back with an **identity response** message, which contains the MS's IMEI. The received IMEI is compared with the IMEI stored in the EIR.

Any time during the update location procedure, MSC/VLR assigns a new TMSI to the MS for security reasons, using the **TMSI reallocation command**. The TMSI can also be assigned at the end within a location update accepted message from the MSC/VLR. In any case, the MS acknowledges receipt of the new TMSI by sending a **TMSI reallocation complete** message to the MSC/VLR.

The MSC/VLR concludes the location update procedure by sending **location update accepted** message transparently to the MS. The VLR is now updated with the new LAI.

The network (MSC/VLR) initiates the release of the control channel by sending a **clear command** message to the BSC. The BSC then instructs the BTS to release the channel by sending a **channel release** message in a **data request** frame. The BTS passes this message to the MS. The BSC also requests the BTS to stop sending SACCH messages by sending **deactivate SACCH** message. The MS, on receiving the **channel release** message from the BTS, confirms the release by sending a LAPDm **disconnect** message. The BTS sends a **release indication** message to the BSC. After link disconnection is achieved, the RF channel is released on instructions from the BSC using RF channel release message. On receiving acknowledgment from the BSC with the **RF channel release acknowledge message**, the BSC informs the MSC, using a **clear complete** message.

3.6.2   Call establishment

The call can be established only if the MS is ON and successfully registered in a network by the location update procedure, as explained in the

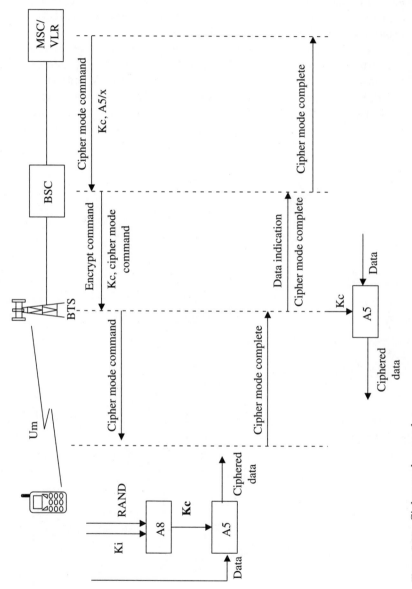

Figure 3-22 Cipher mode procedure.

previous section. There are two different scenarios for call establishment, i.e., mobile originated call (MOC) and mobile terminated call (MTC).

Mobile originated call.   Figures 3-23, 3.24, and 3-25 show the signaling flow within the BSS and the NSS for a mobile originated call. In this example, the called party is a fixed line subscriber.

When a mobile subscriber keys in the destination number, using the keypad and pressing the SEND/OK button, the MS attempts to establish a radio connection with the BTS by sending a **channel request** on a RACH. The channel request message contains a parameter indicating to the BTS the reason for the channel request, i.e., MOC in this case. The BTS, on receiving the channel request message, adds some

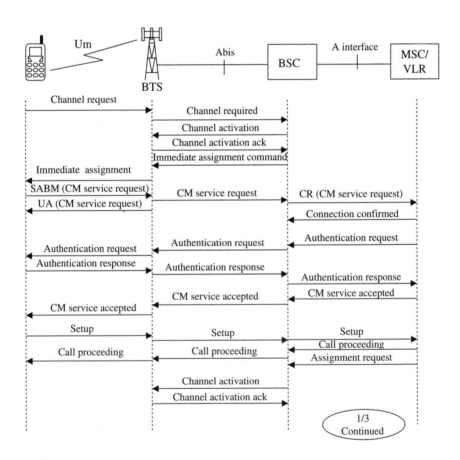

Figure 3-23   MOC-BSS procedures.

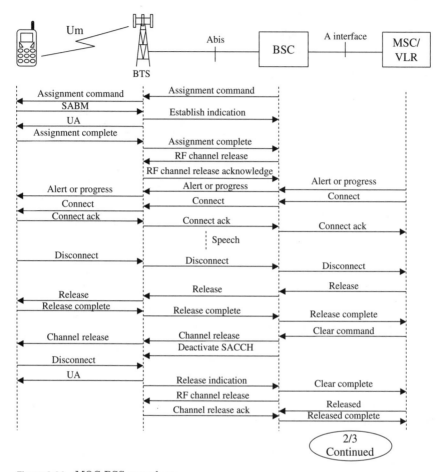

Figure 3-24 MOC-BSS procedures.

radio-related information and sends a **channel required** message to the BSC. The BSC instructs the BTS to reserve a channel by sending the **channel activation** message. The BTS reserves the channel and sends a **channel activation acknowledge** message to the BSC. The BSC then activates the previously reserved channel by sending **immediate assignment command** to the BTS, which passes this message, using AGCH, to the MS. Now the MS initiates establishment of a Layer 2 connection by sending a LAPDm **SABM** frame, which also contains a **CM service request**.

The MS identifies itself by either IMSI or TMSI, which are parameters within the CM service request message. The BTS confirms the Layer 2 connection by a **UA** frame. At the same time, the BTS passes a

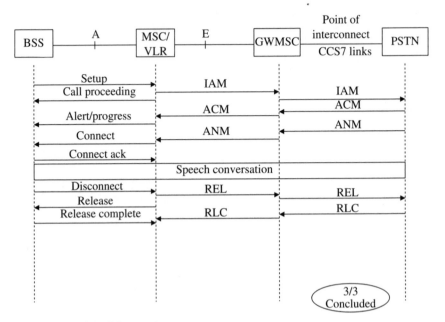

Figure 3-25 MOC-NSS procedures.

CM service request to the BSC. The BSC establishes an SCCP connection to the MSC by sending a **connection request**. This message also contains the CM service request received by the BTS. However, the BSC adds a few additional parameters such as LAC and CI.

The MSC confirms the logical connection by sending an SCCP **connection confirmed** message. The MSC may decide to authenticate the MS at this time. It sends the **authentication request** message over the established SCCP connection to the BSC, which transparently passes this to the MS via the BTS. The authentication request message contains two important parameters: a 128-bit random number (RAND) and a 3-bit ciphering key sequence number (CKSN). The SIM within MS uses Ki, which is stored in the SIM, and RAND to calculate the signed response (SRES) according to the A3 algorithm. The MS sends the SRES value as a parameter within the **authentication response** message to the MSC/VLR.

The VLR compares the SRES received from the MS to the SRES value available with it as a result of the previous send authentication procedure with the HLR/AuC. The MS is successfully authenticated if the two values match.

In networks where ciphering is not enabled, the MSC sends a **CM service accept** message to the MS. If ciphering is enabled, the ciphering mode procedure is initiated by the MSC.

In networks where IMEI check is ON, the MSC/VLR invokes the identity request procedure, as described in the previous section.

For security purposes, the MSC may assign a new TMSI to MS by using the TMSI reallocation command.

The MS now sends a **setup** message transparently to the MSC via the BTS and BSC, using the previously established logical connection. The MSC processes the information contained in the setup message. The most important parameter in the setup message received from the MS is the called party number. The MSC analyses the called party number, creates an ISUP **IAM** message, and sends it to the destination switch in order to establish the connection. The **call proceeding** indication, which is a response from the destination exchange on receiving IAM, is passed back to the MS transparently.

It should be noted that no speech channel is assigned on the radio interface for this call so far. This is to save the radio resources. Now only the MSC triggers the process to assign speech channel for this call. The MSC informs the BSC of the speech channel that is to be used on the A interface, using the **assigned request** command. The BSC then requests the BTS to reserve a traffic channel (TCH) on the air interface, using the **channel activation** message. On receiving a **channel activation acknowledgment** from the BTS, the BSC sends an **assignment command** to assign traffic channel. The MS and the BTS establish Layer 2 connection for the assigned channel by exchanging LAPDm **SABM** and **UA** frames. The BTS establishes Layer 3 connection by sending an **assignment complete** message. As the previously established control channel is no more required, the BSC releases the channel by sending a **channel release** message.

On receiving the ISUP **ACM** message from the destination exchange, indicating the a connection is being set up to the called party, the MSC sends an alert message to the MS. This results in a call progress tone being fed to the MS. The destination exchange sends an ISUP **ANM** message to the MSC when the called party answers. The MSC sends a **connect** message to the MS transparently. On receiving a **connect acknowledgment**, the MSC initiates charging the MS for the call.

The call is released by the MSC on receiving either an ISUP **REL** message from the destination exchange or a **disconnect** message from the MS. The MSC releases all the resources as indicated in the procedures.

Mobile terminated call.   In the example shown here, the call is originated from a fixed line network. The procedures and call flow will essentially remain the same if the call is originated from another PLMN or even from the MS belonging to the same PLMN. The call is routed to the GMSC through a national network point of interconnects (POI) or through an international gateway on the basis of MSISDN, which is an international number and uniquely identifies a PLMN subscriber.

The main task of the NSS, as shown in Figure 3-26, is to find the location of the called party (current serving MSC/VLR) and the temporary identity assigned to the called MS, i.e., MSRN. Once these are known, the incoming call can be routed. The GWMSC has no means to find out where the called MS is. It needs help from the HLR, which is constantly getting updated on MS location as a result of the update location procedure. As shown in Figure 3-26, the GWMSC requests the HLR to get this information using a **send routing info** message. The MSISDN is used to identify the called subscriber. The HLR searches for a MSISDN entry in its record, retrieves the address of the current serving MSC/VLR and the address of the IMSI. The HLR then initiates the **provide roaming number** procedure toward VLR. The VLR assigns a temporary number, i.e., MSRN for routing purposes and provides this number to

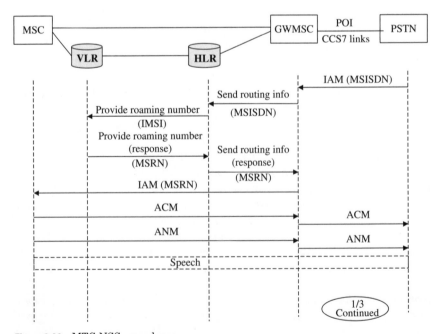

Figure 3-26   MTC-NSS procedures.

the requesting HLR. On receiving the MSRN from the VLR in the provide roaming number response message, the HLR forwards the MSRN to the GWMSC in a response message to the previously received send routing info. Now the GWMSC has all the necessary information to route the incoming call to the called MS. It sends an ISUP **IAM** message to the serving MSC. The MSRN received from the VLR is passed as the called party number parameter within the IAM message (Figure 3-26).

When the MSC receives an IAM from the GWMSC, it queries the VLR to get the location area identity (LAI) of where the MS is currently roaming and invokes the paging procedure (Figure 3-27).

On receiving the **paging** request, the MS requests the BSC to assign a control channel. The call flow (see Figures 3-27 and 3-28) from this point onwards is same as described in the previous section.

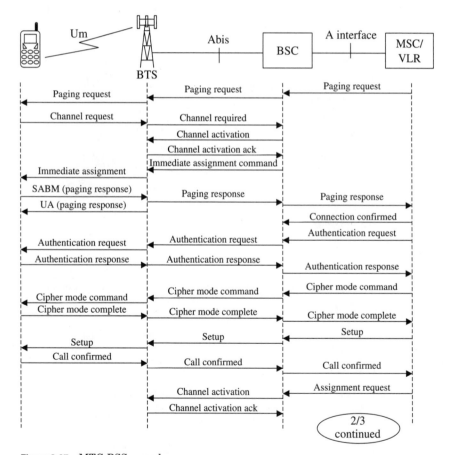

Figure 3-27   MTC-BSS procedures.

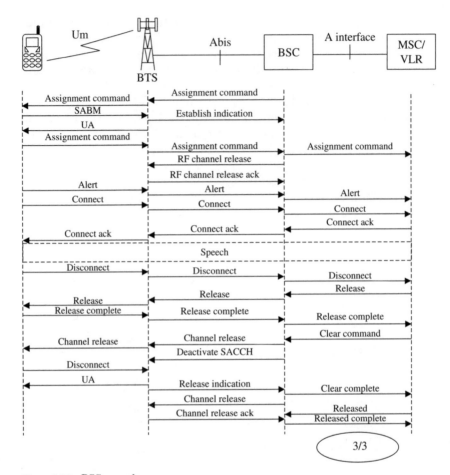

Figure 3-28 BSS procedures.

Bibliography

ITU-T Recommendation E.164, The international public telecommunication numbering plan.

ITU-T Recommendation E.212, The international identification plan for mobile terminals and mobile users.

ITU-T Recommendation E.213, Telephone and ISDN numbering plan for land Mobile Stations in public land mobile networks (PLMN).

ETS 300 102-1 (1990), Integrated Services Digital Network (ISDN); User-network interface layer 3 specifications for basic call control.

3GPP TS 22.001, Digital cellular telecommunications system (Phase 2+); Principles of telecommunication services supported by a Public Land Mobile Network (PLMN).

3GPP TS 29.002, Mobile Application Part (MAP) specification.

ETSI GSM 08.08 Mobile-Services Switching Centre–Base Station System (MSC-BSC) interface; layer 3 specification.

ETSI GSM 08.56 BSC-BTS layer 2 specification.

ETSI GSM 08.58 BSC-BTS layer 3 specification.

4

General Packet Radio Service

4.1 GPRS Overview

GSM offers circuit-switched data services at lower rates. As a standard form, it is limited to 9.6 Kbps. This is obviously not enough for many applications. It takes a long time to setup and is too slow. A user occupies one timeslot out of the available 8 timeslots in a frame during the entire duration of a data call. This makes it very expensive. Later, a few enhancements were proposed and implemented to overcome data rate limitations. High speed circuit-switched data (HSCSD), which uses multiple timeslots, offers a maximum data rate up to 57.6 Kbps. It was not a great success. It essentially a circuit-switched technology and hence very expensive and inefficient for data applications. Most of the data applications are bursty and asymmetric in nature. This means that there are periods when little or no data is being transferred. For data applications, packet switching makes more efficient use of network resources than circuit switching, as it allows sharing of a data channel by many users.

General Packet Radio Service (GPRS) is designed to offer high-capacity end-to-end IP packet services over the GSM infrastructure. It is designed as an overlay network to protect the investment already made in GSM networks. It is based on packet switching. This means that scarce radio resources are shared among many users, resulting in much better utilization and hence lower cost. To achieve high data rates, GPRS employs new air interface error coding schemes and multiple timeslots. By using eight timeslots, the maximum data rate of 171.2 Kbps is achieved. In the BSS, the data is processed separately and passed to new serving nodes capable of handling packet data and finally routed to external data networks such as X.25, the Internet, or intranets.

4.2 GPRS Services

GPRS is designed to address the needs of the mobile data market, which is growing at a fast pace. There are many applications available both for businesses and general users. Generally, these applications are divided into two high-level categories, i.e., horizontal and vertical applications.

4.2.1 Horizontal applications

Horizontal applications are designed for businesses and general consumers and are not specific to any business segment. The nature of these applications suggests an early and huge acceptance but low revenues for the service providers. A few examples of horizontal applications are:

- Intranet access
- Internet access
- Email
- Document and data sharing
- Entertainment
- Marketing
- E-commerce

4.2.2 Vertical applications

Vertical applications are designed specifically for certain business segments. These applications are of high value with identifiable business benefits. The revenue potential for wireless service providers is high. A few examples of vertical applications are:

- Vending, lottery, and ticketing machines
- Vehicle tracking
- Financial services
- Electronic maps
- Telemetry
- Dispatch operations—taxis, field services, delivery services
- Police operations
- Forestry

4.3 GPRS Network Architecture

The GPRS network is designed as an overlay network on an existing GSM network. Additional components are added to handle packet data and interface to external PDNs.

Figure 4-1 shows the additional components required to implement GPRS. It also shows the interfaces between new components, as well as the existing GSM components.

A new mobile station is required to support GPRS. The GPRS terminals are available in many forms and are backward compatible to support GSM voice and circuit-switched data.

The existing GSM BTSs need software upgrades to support a new air interface, new coding schemes, and logical channels and their mapping. No hardware upgrades are required. The BTS connects to the BSC, using the Abis interface as in GSM.

The BSCs require both hardware and software upgrades. The software upgrade is needed to support mobility and paging of GPRS terminals. The hardware upgrade is needed to add new functionality to control and handle the packet data. As shown in Figure 4-1, a packet control unit (PCU) is added to the BSC. The PCU connects with the SGSN node by using the Gb interface, which is based on frame relay.

GPRS introduces two new nodes to handle the packet-switched data. The gateway GPRS support node (GGSN) has capabilities similar to those of the GMSC. It provides an interface (Gi) to external packet data networks (PDNs). The GGSN has mobility management and access server functionality built in.

The serving GPRS support node (SGSN) is an MSC/VLR equivalent. It controls the connection between the MS and the network. The SGSN provides mobility and session management functionality. It connects with the GGSN via the Gn interface. It also has connectivity to HLR, EIR, MSC, and SMS-IWMSC via the Gr, Gf , Gs, and Gd interfaces,

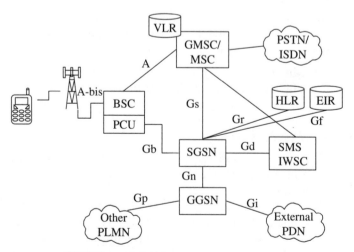

Figure 4-1   GPRS Network Architecture.

respectively. To support inbound roamers, it connects to other PLMNs via the Gp interface.

The GSM HLR needs a software upgrade to support GPRS subscription data and routing information. The HLR communicates with the SGSN via the Gr interface, using the CCS7 MAP protocol. For roaming MSs, the HLR is in a different PLMN than the serving SGSN.

The existing MSC/VLR needs a software upgrade to support terminals that are attached to the both GSM and the GPRS. The MSC/VLR communicates with the SGSN via the Gs interface, using BSSAP+ protocol.

The EIR/AUC does not require any upgrades.

GPRS is also used as an efficient bearer to carry SMS messages. The Gd interface is defined to exchange SMS with the SMS-IWMSC.

The next section describes the new GPRS components and their functionality in more detail.

4.3.1 GPRS terminals

Currently, several different options are available in the market. Some of these have the look and feel of normal mobile phones while others are designed specifically to make better use of enhanced data capabilities. PC cards, Smartphones, and PDAs are very popular GPRS terminals. The ETSI specification defines three different classes of mobiles for the hybrid GPRS/GSM networks:

Class A mobiles can attach to both GSM and GPRS networks simultaneously. These mobiles can make and receive voice and data calls at the same time. In order to achieve this, mobiles monitor both the GSM and the GPRS for incoming calls and have an additional receiver.

Class B terminals can attach to both the GSM and the GPRS networks simultaneously, but can handle only one service at a time. It is possible to switch between the calls. For example, a Class B mobile can suspend an outgoing packet transfer, when it gets an incoming voice call, and resume the packet transfer once the voice call is over.

Class C terminals can attach to only one network, i.e., GSM or GPRS. For example, if a Class C mobile is attached to a GPRS network, it will not be able to make or receive a voice call from a GSM network.

4.3.2 GPRS BSS—Packet control unit

As described earlier, the GSM BSS requires new software for both the BTS and the BSC and additional hardware for the BSC to support GPRS. The new piece of hardware is generally termed a *packet control unit* (PCU). Each BSC will require at least one PCU. One PCU cannot serve multiple BSCs. The PCU connects to the SGSN via a physical and logical data interface, i.e., Gb. In most of the implementations, the PCU is collocated with the BSC, as discussed earlier. However, it is possible that the PCU resides within the BTS or outside the BSC near the SGSN. Figure 4-2 illustrates the three possible locations. The channel control unit (CCU), as shown in Figure 4-2, resides in the BTS and is responsible for channel coding, radio channel measurement, and management functions. It is a software-only implementation.

The Gb interface connects the BSC to the SGSN. This is based on frame relay on the E1/T1 interface. To achieve efficient use of transmission bandwidth, a switched frame relay network is used between the BSC and the SGSN. The newer implementation deploys Gb over IP.

4.3.3   GPRS support nodes

SGSN. The serving GPRS support node (SGSN) provides packet routing to and from the mobile stations currently in its coverage area. For establishing data calls, the GPRS users need to attach to the SGSN via the base station. The SGSN performs functions to support mobility, session, and security management. The SGSN is also responsible for charging functions. To perform its tasks, it communicates with other subsystems using G-interfaces as shown in Table 4-1.

Currently, several vendors supply SGSN with varying performance and capacities. Some of the network providers prefer several SGSNs of smaller capacity, while others like to consolidate and implement only a few

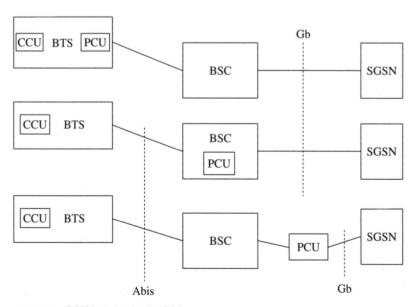

Figure 4-2   PCU location in the BSS.

TABLE 4-1 SGSN Interfaces

Connection	Mandatory/ optional	Interface	Common implementation
SGSN-PCU	Mandatory	Gb	Frame relay–based, E1/T1 interface, channelized, nonchannelized, and fractional. In most of the implementations, each Gb link consists of $n \times 64$ timeslots, depending on traffic.
SGSN-HLR	Mandatory	Gr	CCS7-based, E1/T1 interface. Multiple timeslots can be used for signaling, if required
SGSN-EIR	Optional	Gf	CCS7-based, E1/T1 interface. Multiple timeslots can be used for signaling, if required
SGSN-MSC/ VLR	Optional	Gs	CCS7-based, E1/T1 interface. Multiple timeslots can be used for signaling, if required
SGSN-SMS GMSC SGSN- SMS IWMSC	Optional	Gd	CCS7-based, E1/T1 interface. Multiple timeslots can be used for signaling, if required
SGSN-GGSN	Mandatory	Gn	IP-based IP over Ethernet/Fast Ethernet IP over ATM IP over PPP
SGSN-external GSNs	Optional	Gp	IP-based IP over Ethernet/Fast Ethernet IP over ATM IP over PPP

higher-capacity SGSNs covering the whole network. The SGSN performance and capacity are defined by the following parameters:

- Maximum number of simultaneously attached users
- Maximum number of PDP contexts
- Maximum throughput

PDP stands for packet data protocol. In the GPRS context, it is X.25 or IP.

In most of the implementations, the SGSN and the GGSN functions reside in separate physical nodes. However, it is possible to combine the SGSN and the GGSN functionalities in a single node. In such cases, the Gn interface is not visible.

Mobility management. Like an MSC in GSM, SGSN is responsible for supporting MS mobility. It keeps track of all the subscribers in its coverage area. The MM functions include the following procedures:

- GPRS attach
- GPRS detach
- Paging
- Routing area update

These procedures are discussed later in this chapter.

Session management. The SGSN is responsible for relaying the data PDUs between the MS and the GGSN. For this purpose, SGSN establishes a session, which is defined as the period between opening and closing the connection. The SM functions include the following procedures:

- Activate PDP context
- Modify PDP context
- Delete PDP context

These procedures are discussed later in this chapter.

Security management. The SGSN authenticates the subscriber at the very first request to attach to the network. This is necessary to prevent unauthorized users from gaining access to the network services.

As the air interface is most vulnerable to fraudulent access, the SGSN also initiates procedures with the MS to cipher the data. The ciphering, however, is a network feature, which can be put off if the network operator so desires. The security functions include following procedures.

- Identity request
- Authentication and ciphering

PDU handling. The SGSN and the GGSN use this function to transport packet data units (PDUs) between the MS and the external packet data network. The SGSN and the GGSN use a tunneling concept to transport PDUs over the Gn interface. The PDUs are encapsulated into an IP datagram to facilitate transfer of PDUs of any format across the Gn link.

Charging. The charging functions include the call detailed record (CDR) generation and charging gateway function (CGF). The CDR includes information necessary for the service providers to invoice the customers. The CDRs are generated for each PDP context and contain parameters such as user identity, PDP address, volume of data transfer, time, and QoS requested and assigned. The CGF functions include collection, temporary storage, and transportation of CDRs to downstream billing system.

Operation and maintenance. O&M functions are vendor specific. They allow the service provider to manage the nodes. Generally, all the FCAPS (fault, configuration, accounting, performance, and security) functions are supported.

GGSN. As the name suggests, the gateway GPRS serving node acts as a gateway between the GPRS network and the external PDN. Several SGSNs can use one GGSN to access an external PDN. On the other hand, a SGSN can send its packets by using different GGSNs to reach different packet data networks. The GGSN functions include session management, PDU handling, PDP address management, QoS negotiation, and authentication through RADIUS. To perform its tasks, it communicates with other subsystems using G interfaces as shown in Table 4-2.

Session management. The GGSN works in conjunction with the SGSN to establish a session. The GGSN is also responsible for assigning an IP address to the MS. It supports both dynamic and static IP address allocation. For dynamic address allocation, the GGSN either provides an IP address from its own allocated IP address range or uses a RADIUS server located at the ISP or a company's intranet.

Mobility management. The GGSN makes sure that the PDUs related to an MS get tunneled to its current serving SGSN.

Interface to external PDN and PDU handling. GGSNs act as an interface point to external networks such as the Internet, enterprise intranets, ISPs, and to other GPRS PLMNs. To external PDNs, the GPRS network is another data network. It hides GPRS network complexity such as mobility from the external networks and acts as a router for the IP addresses of all the MSs served by the GPRS network.

Security. The GGSNs include a firewall to protect the GPRS network from intrusion. The GGSNs also include a RADIUS client, which queries an external RADIUS server on behalf of MS for authentication purposes.

TABLE 4-2   GGSN Interfaces

Connection	Mandatory/ optional	Interface	Common implementation
GGSN-SGSN	Mandatory	Gn	IP-based IP over Ethernet/Fast Ethernet IP over ATM IP over PPP
GGSN-HLR	Optional	Gc	CCS7-based, E1/T1 interface. Multiple timeslots are used for signaling, if required.
GGSN-external PDN	Mandatory	Gi	IP-based IP over Ethernet/Fast Ethernet IP over ATM IP over PPP

Charging. Like SGSNs, the GGSNs also have charging and CGF functions. The CDRs from both the SGSN and the GGSN are needed to support billing and invoicing functions. For example, the GGSN is unaware of MS location; hence the CDRs available with GGSNs are not sufficient for roaming charging.

4.4 Interfaces and Protocols

4.4.1 User plane

Figure 4-3 shows the protocol architecture of the GPRS user/transmission plane. The user plane provides user information transfer and associated signaling procedures, e.g., flow control and error detection and correction.

GGSN-SGSN. The user data packets arriving at the SGSN from the MS or at the GGSN from the external PDN are encapsulated before onward transmission within the GPRS backbone network. The GPRS tunneling protocol for user plane (GTP-U) is used to tunnel the user data between SGSN and GGSN over the Gn interface and between GSNs from different PLMNs over the Gp interface. GTP carries the user packets, i.e., X.25 or IP.

TCP/UDP is used to transport GTP packets within the GPRS intra-PLMN backbone. TCP carries GTP PDUs (G-PDUs) for protocols that require a reliable data link, e.g., X.25. UDP carries G-PDUs for protocols that do not require a reliable data link, e.g., IP.

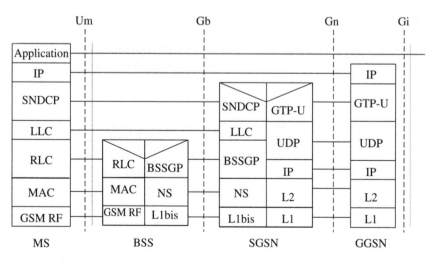

Figure 4-3 Protocol architecture of the GPRS user/transmission plane.

Below UDP/TCP, IP is used as a network layer protocol to route the packets from the upper layer through the backbone network. Currently, IPv4 is used with a future option to upgrade to IPv6.

Figure 4-4 shows the encapsulation of user data at the GGSN for further tunneling through the intra-PLMN backbone network. Note that each layer adds its own overheads. The resulting PDU at the network layer, which is transferred over the Gn interface, is called N-PDU.

SGSN-BSS. As shown in Figure 4-3, the subnetwork dependent convergence protocol (SNDCP) is used to transfer data packets between SGSN and MS. SNDCP is designed to carry N-PDU transparently between SGSN and MS regardless of the network layer protocol, i.e., IP, X.25, or any other future protocol that an end application might use. SNDCP also converts the network layer PDUs on the Gn interface into a format suitable for the underlying GPRS network architecture. The functions of SNDCP include the following:

- Multiplexing of N-PDUs from one or several network layer entities (PDPs such as X.25 or IP) onto a virtual logical connection

- Buffering of PDUs for acknowledge service

- Delivery sequence management for each NSAPI (see Section 4.5 for the definition)

- Compression and decompression of the user data

- Compression and decompression of protocol headers

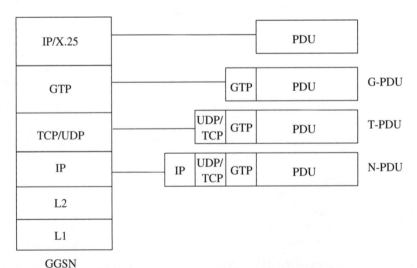

Figure 4-4 User data packet transport across GPRS backbone.

- Segmentation of a network protocol data unit (N-PDU) into LLC protocol data units (LL-PDUs) and reassembly of LL-PDUs into an N-PDU
- Negotiation of the control parameters between the SNDCP entities

Below SNDCP, logical link control (LLC) protocol is used for packet data transfer between the SGSN and the MS. The LLC provides a highly reliable, ciphered logical link between the MS and the SGSN. The LLC frame format is based on the LAPD protocol with a few modifications to make it suitable to be used on a radio link. It uses both acknowledged and unacknowledged data transfer, depending upon the requirement on QoS. The LLC also manages frame retransmission and buffering based on the negotiated QoS.

The data from several mobile stations is multiplexed over a Gb link in downlink direction. The same is true for the uplink direction, where the data destined to several MSs is to be multiplexed over a Gb link. How can LLC frames belonging to a MS be routed to the right MS (RLC/MAC) via a BSS? The base station subsystem GPRS protocol (BSSGP), which is a new and GPRS-specific protocol, in conjunction with the network service (NS) layer, performs this task.

The tasks performed by the BSSGP are:

- In the uplink direction, the BSSGP at the BSS provides the needed information to route the user data to the SGSN. The information is derived from the RLC/MAC.
- In the downlink direction, the BSSGP layer at the SGSN provides radio-related information used by the RLC/MAC function.
- Node management functions between the SGSN and the BSS.

The relay function at the BSS transfers LLC frames between the RLC/MAC layers and the BSSGP layer. The BSSGP uses the following identifiers to indicate to the NS layer the destination of packets:

- BSSGP virtual connection identifier (BVCI)
- Link selection parameter (LSP)
- Network service entity identifier (NSEI)

BVCI identifies entities at the SGSN and the BSS between which the data and signaling information is to be transferred. Each BVCI between two peer entities is unique.

In the case of load sharing, LSP is used in conjunction with the BVCI to identify a physical link. The BSSGP virtual connection between the SGSN and the BSS is uniquely identified with the combination of BVCI and NSEI.

The network service (NS) layer uses frame relay over the Gb interface. The NS layer uses a data link connection identifier (DLCI) to indicate the routing path between SGSN and the BSS. The NS layer derives the DLCI value from the BVCI, LSP, and NSEI given by BSSGP Layer.

Figure 4-5 shows the concept of an end-to-end BSSGP virtual channel between the BSS and the SGSN. A summary of addressing over Gb is as follows:

- The physical connection (bearer) between the BSS and the SGSN is E1/T1. The bearer channel (BC) carries frame relay signaling and data.

- On a bearer channel, several logical flows are maintained; i.e., permanent virtual connections (PVCs). The PVC is identified by the DLCI. These are set by the network providers.

- A network service virtual links identifier (NSVLI) identifies the virtual link on a physical bearer. NSVLI = DLCI + BC.

- The end-to-end virtual connection between the BSS and the SGSN is known as NS-VC.

- A group of NS-VCs is identified by NSEI.

BSS-MS. Layer 2, the data link layer, at the Um interface consists of two sublayers:

- A logical link control layer between MS and SGSN, which has been described in the previous section

- A radio link control (RLC)/medium access control (MAC) layer

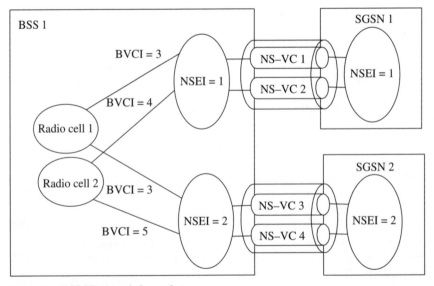

Figure 4-5 BSSGP virtual channel.

The main task of the RLC sublayer is to establish a reliable link between the MS and the BSS. The functions of RLC layer include:

- LLC PDU transfer between the LLC and the MAC layers.

- Segmentation of LLC PDUs into smaller RLC data blocks and reassembly of the blocks to fit into a TDMA frame. This is done because the LLC PDU size is too big to be transferred on the air interface efficiently. A unique temporary frame identity (TFI) identifies each segment. The TFI is derived from the MS identifier TLLI (see Section 4.5 for the definition) and the frame sequence number.

- Backward error correction of RLC data blocks. The backward error correction is based on the NAK automatic repeat request (ARQ) protocol. If the receiving RLC entity detects a missing TFI, it requests retransmission of the missing block. Once the missing block is available, the LLC frame is built and passed to the upper layer.

The medium access control (MAC) controls and manages the common transmission medium to enable data transfer from and to multiple MSs. It employs algorithms for contention resolution, scheduling, and prioritization based on negotiated QoS.

The physical interface between the MS and the BSS is divided into two sublayers, i.e., the physical link layer (PLL) and the physical RF layer (RFL). The PLL resides at the physical channels and provides services for channel coding, error detection, and error correction. It is also responsible for interleaving one radio block onto four consecutive bursts. The RFL is responsible for modulation, demodulation, frequency selection, etc.

4.4.2 Signaling plane

The signaling plane architecture consists of a set of protocols to support the functions of the transmission/user plane. Most of the protocols used are the same as those in the transmission plane.

Figure 4-6 illustrates the layered protocol architecture for the signaling between the MS and the SGSN. The Layer 3 protocols and their functions are as follows.

GPRS mobility management (GMM) supports mobility management functions. GMM includes functions such as GPRS attach/detach, security, and cell and routing area update.

Session management (SM) includes the function to create, manage and control the user sessions. Create PDP context, Delete PDP context are a few examples of session management procedures.

GMM/SM and short message service (SMS) rely on the LLC layer to transfer messages between the MS and the SGSN. GMM/SM layers

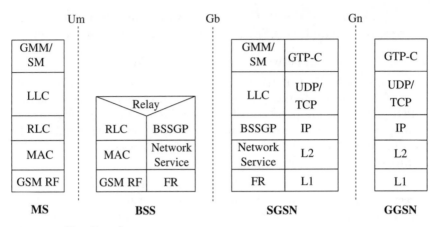

Figure 4-6 Signaling plane.

are transparent to the underlying layers between the MS and the BSS. This means GMM/SM messages are transported between the MS and the SGSN transparently through the BSS.

MAP protocol is used between the SGSN and the HLR (Figure 4-7). It has been extended to support GPRS-specific procedures. The applications/users, i.e., SGSN and HLR, use the MAP protocol to transport

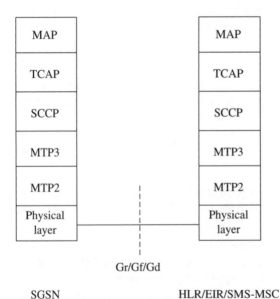

Figure 4-7 SGSN-HLR signaling.

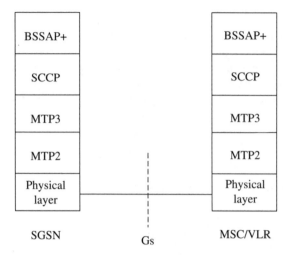

Figure 4-8 SGSN-MSC/VLR signaling.

the signaling information related to location update, subscription data, handovers, etc. The MAP protocol is described in Chapter 3. TCAP, SCCP and MTP are CCS7 protocols and are described in Chapter 2.

The GSM base station subsystem application part (BSSAP) has been extended to support GPRS specific procedures and is called BSSAP+. It is used for signaling transfer between the MSC/VLR and the SGSN (Figure 4-8). BSSAP+ supports the procedures for combined GPRS/IMSI attach, combined location update (GSM & GPRS), and paging for an MS using GPRS. BSSAP+ relies on SCCP and the underlying MTP protocol to transport messages between communicating entities.

The GPRS tunneling protocol control plane (GTP-C) is used between two GSNs over the Gn/Gp interface. The GTP-C signaling flow is logically associated with, but separate from, the GTP-U tunnels. This protocol tunnels signaling messages between SGSNs and GGSNs (Gn) and between SGSNs in the backbone network (Gp). This supports procedures such as create PDP context and PDU notification.

4.5 GPRS Identities

International mobile subscriber identity (IMSI) is a unique identifier for a GSM/GPRS subscriber in a PLMN. It is stored in the SIM and also in the HLR as part of the subscriber data. Chapter 3 describes the temporary identities associated with the MS in the GSM networks. Likewise, in GPRS, the MS is assigned temporary identities. These are used at different interfaces and serve specific purposes.

4.5.1 P-TMSI

A packet temporary mobile subscriber identity (P-TMSI) is assigned to a GPRS MS at the time of GPRS attach. Like TMSI, P-TMSI is used to avoid transmitting IMSI over the air interface. P-TMSI is of local significance and is applicable in the area served by an SGSN. If the MS moves out to a new SGSN area, the current serving SGSN assigns a new P-TMSI to the MS.

4.5.2 TLLI

The temporary logical link identity (TLLI) is a temporary identity used during a PDP session over the Um and Gb interfaces. The MS or the SGSN derives the TLLI, using the P-TMSI.

The TLLI can be derived in one of the three ways:

1. A local TLLI is built by using the P-TMSI assigned by the SGSN.

2. A foreign TLLI is derived from a P-TMSI allocated in a different routing area.

3. The MS generates a random TLLI in the absence of a valid P-TMSI.

The network can assign a new P-TMSI any time. The MS then derives the value of TLLI by using the new P-TMSI.

4.5.3 NSAPI

The network layer service access point identifier (NSAPI) is used with TLLI for network layer routing. The NSAPI acts as an index for the appropriate packet data protocol (PDP) that is using the services of SNDCP. When an IP packet is received at the MS for a particular IP address at a service access point, the IP packet is encapsulated and the NSAPI (from the previous activation of PDP context) value is attached. TLLI is set to the MS's TLLI before the encapsulated IP packet is passed to the SNDCP layer. The SGSN, on receiving the IP PDU, analyzes the TLLI and NSAPI and forwards the IP PDU to the right GGSN. The SGSN maintains a table with information similar to that shown in Table 4-3 to make the routing decision.

TABLE 4-3   Example SGSN Network Layer Routing Data

MS1	TLLI = 1 NSAPI 8	TEID 2	GGSN IP
MS2	TLLI = 2 NSAPI 6	TEID 5	GGSN IP
MS3	TLLI = 3 NSAPI 3	TEID 6	GGSN IP

4.5.4 TEID

A tunnel end point identifier (TEID) is used by the GTP protocol between GSNs to identify a tunnel end point in the receiving GTP-C or GTP-U protocol entity and to identify a PDP context. As shown in Table 4-3, each PDP context has a one-to-one relationship between the TEID and the NSAPI/IMSI (IMSI and TLLI have a one-to-one relationship). The receiving-end side of a GTP-U tunnel locally assigns the TEID value. The TEID value is then made known to the transmitting side by GTP-C protocol.

4.6 GPRS Procedures

4.6.1 Mobility management

Mobility management states. The mobility management (MM) function is to support the mobility of user terminals. The MM activities related to a GPRS terminal are characterized by one of the three different states, i.e., IDLE, STANDBY, and READY. Figures 4-9 and 4-10 illustrate the MM state models of the MS and the SGSN, respectively.

In GPRS IDLE state, the MS camps onto the GSM network. The MS can receive circuit-switched paging and perform location area updates. In this stage, the MS behaves like any other GSM phone. It is not attached to GPRS mobility management yet. The MS and the SGSN contexts hold no valid location or routing information for the subscriber.

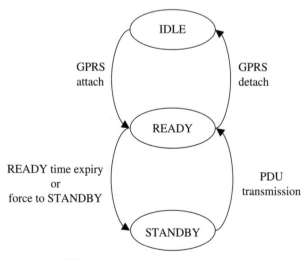

Figure 4-9 MM state model of an MS.

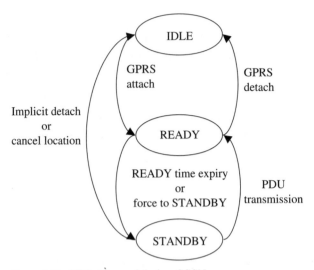

Figure 4-10   MM state model of an SGSN.

Data transmission to and from the MS is not possible. The GPRS MS is not reachable, as paging is not possible. The MS makes the transition to the STANDBY state by attaching itself with the network using the GPRS attach procedure.

In the GPRS READY state, the network is aware of cell location as a result of the successful mobility management procedure. The MS can send or receive PDP PDUs and activate and deactivate PDP contexts. The MS makes the transition to the STANDBY state if no data transmission occurs for a settable timer period. A GPRS detach procedure brings the MS back to the IDLE state.

In GPRS STANDBY, the MS is attached to the network. The MS and the SGSN have established MM contexts. The MS can be paged for PS and CS call via SGSN. The MS can initiate activate and deactivate PDP context. The MS makes the transition to the READY state by transmitting or receiving an LLC PDU. The transition to the IDLE state happens when the MS implicit detach from the network occurs or the SGSN receives a MAP cancel location from the HLR.

GPRS attach.   The mobile station must perform a GPRS attach in order to be known to the network and move to the READY state. Following the successful GPRS attach, an MM context is said to be active at the MS and the SGSN. The MS can activate PDP context only after a successful attach.

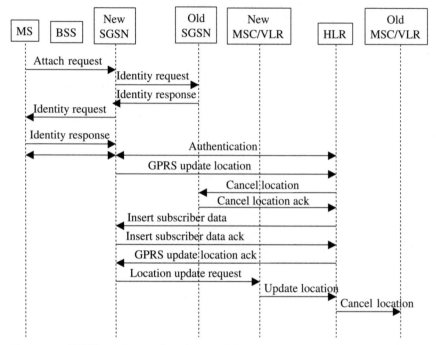

Figure 4-11 GPRS attach procedure (part 1 of 2).

Figures 4-11 and Figure 4-12 show the steps for the GPRS attach procedure.

1. The MS sends the attach request message to the serving SGSN. The key parameters are:
   - IMSI or P-TMSI
   - Routing area identifier (RAI)
   - Cipher key sequence number
   - Attach type
   - DRX

2. If the MS is known to the SGSN, i.e., the SGSN has not changed since the MS was last attached to the network, then step 3 is not required.

3. If this MS is unknown to the serving SGSN, the SGSN takes additional steps to get the IMSI from the old SGSN. In case the old SGSN failed to provide the IMSI, the new SGSN requests the MS to provide the IMSI. The identity request message is used in both cases.

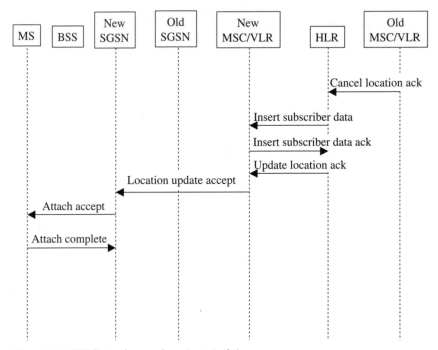

Figure 4-12   GPRS attach procedure (part 2 of 2).

4. Once the IMSI is known, the serving SGSN initiates the authentication procedure by sending an authentication and ciphering request' to the MS. The procedure is described in Section 4.6.3.

5. If the SGSN has changed since the MS was last attached to the network, the new serving SGSN initiates the update location procedure with the HLR.

6. The HLR sends a cancel location to the old SGSN and sends subscription data to the new SGSN by sending an insert subscriber data message.

7. The SGSN sends an attach accept to the MS. The MS acknowledges by sending an attach complete message.

In the case of GPRS attach failure, one of the following errors is returned in the attach reject message.

- Illegal MS
- GPRS service not allowed
- GPRS and non-GPRS services not allowed

- PLMN not allowed
- Location area not allowed
- Roaming not allowed in this location area
- GPRS services not allowed in the PLMN
- No suitable cells in location area

The Gs interface between the SGSN and the MSC/VLR is not a mandatory interface. In case this interface is active, the MS can perform combined GPRS/IMSI attach. In addition to action shown in step 6, the HLR also performs a cancel location with the old MSC/VLR and updates the new MSC/VLR with the subscription data.

GPRS detach.   GPRS detach can be performed either by the MS or the network. The MS uses this procedure to inform the network that it does not want GPRS services anymore. The SGSN or the HLR initiates a detach procedure to inform the MS that GPRS services are no more accessible to the MS. The three different types of detach are:

- IMSI detach
- GPRS detach
- Combined IMSI/GPRS detach

The detach request could be explicit or implicit. The explicit detach can be initiated by the MS or the network.

In case of implicit detach, it is the network that initiates the detach procedure, without informing the MS. This may happen in the following cases: (1) after the mobile reachable timer expiry and (2) once the logical link disconnects because of irrecoverable radio error causes.

Figure 4-13 shows the MS-initiated detach procedure. The information elements included in the detach request message are:

- Detach type (GPRS detach only, IMSI detach only, combined GPRS/IMSI detach)
- P-TMSI
- Switch off (detach because of switch off)

Figure 4-14 shows the SGSN-initiated detach procedure. The main parameter in the detach request message is detach type. This parameter indicates whether the MS has been requested to perform a new attach and PDP context activation for the previously activated PDP contexts. If yes, then the attach procedure will be initiated when the detach procedure is completed.

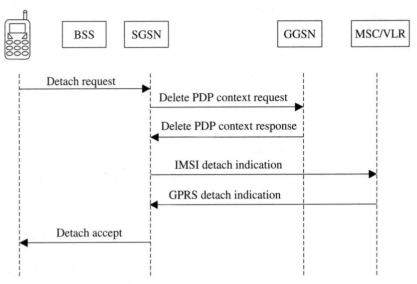

Figure 4-13 MS-initiated detach procedure.

Figure 4-15 shows the HLR-initiated detach procedure. The HLR initiates this procedure as a result of operator action to invoke barring or withdrawing the subscription.

Paging. Figure 4-16 shows the GPRS paging procedure. On receiving a PDU from GGSN/SGSN meant for a MS in its serving area, the SGSN

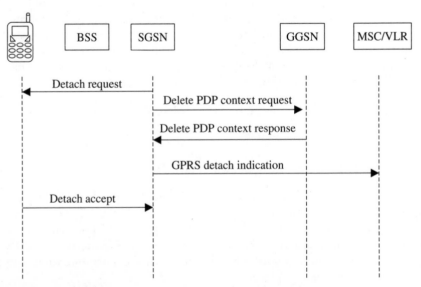

Figure 4-14 SGSN-initiated detach procedure.

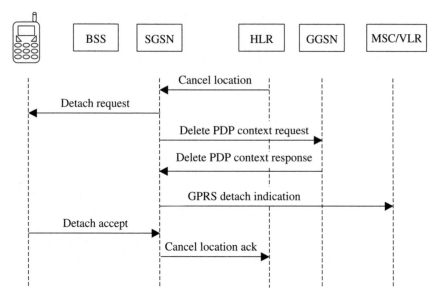

Figure 4-15 HLR-initiated detach procedure.

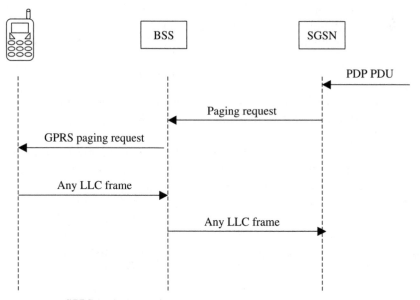

Figure 4-16 GPRS paging procedure.

sends a BSSGP paging request to the serving BSS. The main parameters included in this request are:

- IMSI
- P-TMSI
- Routing area
- Channel needed (indicating GPRS paging)
- Negotiated QoS
- DRX parameters

The BSS checks for the routing area in the paging request message and pages the MS in each cell belonging to RA. The MS responds with any LLC frame, e.g., RR or info frame, other than NULL LLC. The BSS, on receiving the LLC frame, adds cell global identity (CGI) and sends the LLC frame to the SGSN. The SGSN considers the LLC frame a successful paging response.

4.6.2 Session management

MS-initiated PDP context activation.   Once the MS is attached to the GPRS network, it can send and receive SMS. The MS must perform PDP context activation to use other GPRS services such as Internet access, intranet access, email, and MMS. This is required to establish a tunnel between the MS and the requested external packet data network for the data transfer. Figure 4-17 illustrates the PDP context activation procedure initiated by the MS.

The steps to successful PDP context activation are as follows:

1. The MS sends an activate PDP context message to the serving SGSN. This message contains following parameters.
   - PDP type (IP or X.25)
   - PDP address (static IP address or NULL for dynamic IP address)
   - APN (access point name: points to a certain packet data network or service a user wishes to access)
   - QoS requested
   - NSAPI
   - PDP configuration options
2. The SGSN may decide to perform standard security checks, i.e., ciphering and authentication, IMSI check, IMEI check, P-TMSI reallocation, etc.)
3. The SGSN validates the activated DP context request for PDP type, PDP address, APN, etc. against the subscription. The SGSN also requests its local DNS to provide the GGSN address serving the

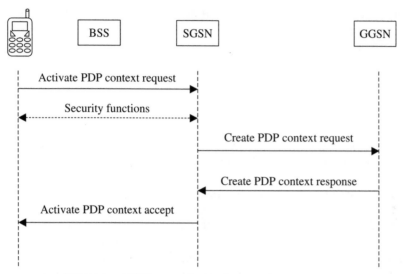

Figure 4-17 MS-initiated PDP context activation procedure.

requested APN. If any of the validation checks fail, the SGSN rejects the request and provides an appropriate cause value. On successful validation, the SGSN determines the tunnel ID (TID) by a combination of IMSI and NSAPI and sends a create PDP context request message to the GGSN. This message contains the following parameters.

- PDP type (IP or X.25)
- PDP address (static IP address or NULL for dynamic IP address)
- APN (access point name: points to a certain packet data network or service a user wishes to access)
- QoS negotiated
- TID
- NSAPI
- MSISDN
- Selection mode (subscribed or non subscribed APN)
- PDP configuration options

4. The GGSN uses APN to identify the packet data network or services using DNS. It also uses DHCP or an external RADIUS server to get a PDP address for the MS. If the GGSN has been configured to use external PDN address allocation for the requested APN, the PDP address is set to 0.0.0.0, indicating that the PDP address shall be negotiated by the MS with the external PDN after the PDP context is activated.

5. The GGSN sends the create PDP context response to the SGSN. This message contains the following parameters.
   - PDP address
   - QoS negotiated
   - TID
   - PDP configuration options
   - BB protocol (TCP/UDP)
   - Cause

6. The SGSN inserts address parameters, i.e., NSAPI and GGSN address and sends an activate PDP context response message to the MS.

Network-initiated PDP context activation.   When a GGSN receives a PDP PDU, it checks whether a PDP context exists for the PDP address. If not, the GGSN tries to deliver the PDP PDU by initiating a network-initiated PDP context request. Figure 4-18 illustrates this procedure. Network-initiated PDP context activation is possible only if the GGSN has static PDP information about the PDP address. The steps to successful PDP context activation are as follows:

1. On receiving a PDP PDU, the GGSN checks if there is static PDP information for that PDP address. If so, it starts storing subsequent PDP

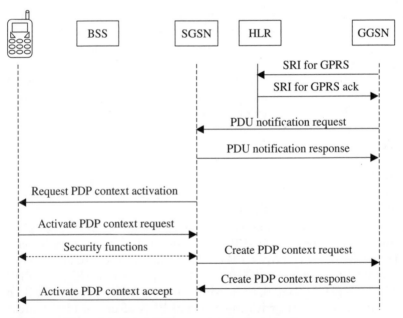

Figure 4-18 Network-initiated PDP context activation procedure.

PDUs for that PDP address. It sends a send routing information for GPRS message to HLR. The HLR returns a send routing information for GPRS ack message with the following parameters:

- IMSI
- SGSN address

In cases where the request cannot be served, the HLR returns a negative acknowledgement with appropriate reason (e.g., IMSI unknown in the HLR).

2. The GGSN sends a PDU notification request to the SGSN. The message contains the following parameters:

- IMSI
- PDP type
- PDP address
- APN

The SGSN returns a PDP notification response, indicating to the GGSN that it will request the MS to activate the PDP context.

3. The SGSN sends a request PDP context activation message to the MS with the following parameters:

- PDP type
- PDP address
- APN

4. The MS then initiates a PDP context activation procedure as defined in the previous section.

PDP context modification.   By using this procedure, a previously negotiated PDP context can be modified on request from the MS, SGSN, or GGSN. The parameters, which can be modified, are as follows:

- QoS negotiated
- Radio priority
- Packet flow ID

In addition to these, the GGSN can also request a PDP address change. Figure 4-19 illustrates the GGSN-initiated PDP context modification procedure.

The steps are as follows;

1. The GGSN sends an update PDP context request message to the SGSN. In this case, assume that the modification request is to change the previously negotiated QoS profile.

2. The SGSN checks the requested QoS profile against its capabilities, current load, and subscribed QoS profile. The SGSN then selects

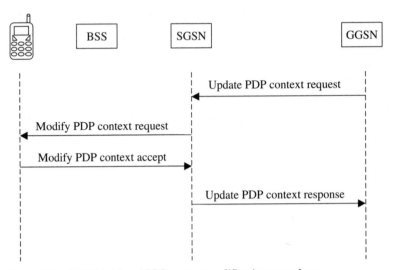

Figure 4-19 GGSN-initiated PDP context modification procedure.

radio priority and packet flow ID on the basis of the negotiated QoS profile and sends a modify PDP context request message to the MS.

3. The MS checks if it can accept the request. If yes, it sends a modify PDP context accept message to the SGSN. If no, it initiates deactivate PDP context procedure with the SGSN.

4. On receiving the modify PDP context accept message, the SGSN returns an update PDP context response message to the GGSN.

5. In cases where MS initiates the deactivate PDP context procedure, the SGSN follows the deactivation procedure.

PDP context deactivation.   The MS, SGSN, or GGSN can initiate the PDP context deactivation procedure. Figure 4-20 illustrates the MS-initiated PDP context deactivation procedure.

1. The MS sends a deactivate PDP context message to the SGSN. The message contains a teardown indication.

2. The SGSN sends a delete PDP context request message to the GGSN. The message contains TEID, NSAPI, and a teardown indication.

3. The GGSN removes all the PDP contexts associated with the PDP address and returns a delete PDP context response message to the SGSN.

4. The SGSN returns a deactivate PDP context accept message to the MS.

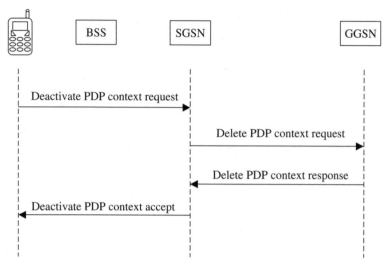

Figure 4-20   Delete PDP context procedure.

4.6.3   Security function

The objectives of the security functions are to prevent an unauthorized user to access the services and keep user identity confidential.

Authentication procedure.   The authentication procedure, as illustrated in Figure 4-21, is similar to the procedure used in GSM. The SGSN initiates the authentication procedure when a GPRS MS tries to attach to the network, using the GPRS attach procedure.

1. In cases where the SGSN has previously stored authentication triplet, then steps 1 and 2 are not required.

2. In cases where the SGSN does not have a previously stored authentication triplet, it requests the HLR to provide authentication triplet by sending a send authentication info message.

3. The HLR responds with a send authentication info ack message. This message has triplets consisting of RAND, SRES, and Kc.
   - RAND: random access number
   - SRES: signed response
   - Kc: ciphering key

4. The SGSN then sends an authentication and ciphering request with following information elements.
   - RAND
   - CKSN: ciphering key sequence number
   - Ciphering algorithm, i.e., A5

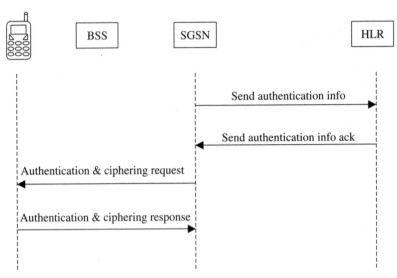

Figure 4-21 Authentication procedure.

5. The MS computes the SRES value and sends in its response to the SGSN using an authentication and ciphering response message. The MS then starts ciphering.

6. If the SRES from the MS matches with the one received from the HLR, the user is successfully authenticated.

P-TMSI reallocation procedure. The temporary logical link identity (TLLI) is a temporary identity used during a PDP session over the Um and Gb interfaces. The MS or the SGSN derives the value of TLLI by using P-TMSI. Only the MS and the SGSN are aware of the relationship between the IMSI and the TLLI. The SGSN may reallocate P-TMSI any time. The MS is forced to compute a new TLLI. The P-TMSI reallocation procedure, as illustrated in Figure 4-22, is initiated by the SGSN any time, or it can be included in the GPRS attach or RA update procedure.

1. The SGSN sends a P-TMSI reallocation command to the MS. The message contains the following information elements:
   - New P-TMSI
   - P-TMSI signature (optional)
   - RAI

2. The MS responds with a P-TMSI reallocation complete message to the SGSN.

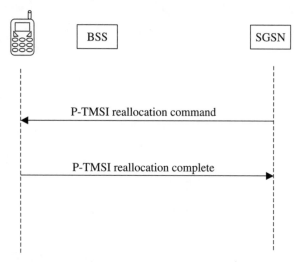

Figure 4-22   P-TMSI reallocation procedure.

Bibliography

3GPP TS 29.002, Mobile Applications Part (MAP) specification.
3GPP TS 23.060, General Packet Radio Service, Service Description, Stage 2.
3GPP TS 23.003, Numbering, addressing and identification.

Chapter

5

Third Generation Networks

"Third generation" (3G) is a new network technology being deployed in many mobile networks today. The motivations to develop a new technology when existing wireless technologies are quite mature and serving users very well are these:

- Demand for high-speed data
- Capacity limitations of existing networks
- Demand for seamless roaming worldwide
- Demand for more mobile data centric services
- Wireless multimedia services
- Convergence between telecommunication, IT, media, and content
- Desire to access data anywhere and anytime
- Seamless service environment—wireless, wireline, home, office, on the move.

5.1 3G Specifications

The International Telecommunication union (ITU) started the efforts to create a common radio interface technology on a global basis under the umbrella IMT-2000 family of standards.

The key objectives set by the ITU for Third Generation networks are:

- Integration of residential, office, and cellular services into a single system
- Enable high-speed data applications, i.e., 144-Kbps data for high-speed vehicular environments, 384-Kbps data for pedestrian or low-speed vehicular environments, and 2-Mbps data for stationary environments

- Unique subscriber number independent of the network and service provider
- Capacity and capability to serve more than 50 percent of the population
- Integration with satellite components
- Seamless roaming
- Quality of service commensurate with that of terrestrial networks at an affordable cost

Several regional standardization bodies have supported the ITU IMT-2000 initiative. The regional standard organizations, such as TIA and T1 in the United States, ETSI in Europe, TTC and ARIB in Japan, CWTS in China, and TTA in Korea submitted their proposals on radio access technologies for open review. The five accepted radio access technologies are:

- IMT-2000 CDMA direct spread (IMT-DS), referred to as UTRA-FDD, UMTS FDD, WCDMA
- IMT-2000 CDMA multicarrier (IMT-MC), referred to as CDMA2000
- IMT-2000 CDMA TDD (IMT-TC), referred to as UTRA-TDD, and in China as SCDMA
- IMT-2000 CDMA single carrier (IMT-SC), referred to as UWC-136/EDGE
- IMT-2000 FDMA/TDMA (IMT-FT), referred to as DECT

A new 3GPP–Third Generation Partnership Project organization was formed in collaboration with ETSI and other regional standards bodies to define and maintain a Universal Mobile Telecommunication System (UMTS) specification. The current organizational partners are ARIB, CCSA, ETSI, ATIS, TTA, and TTC. UMTS embraces WCDMA as its radio technology. WCDMA is the dominating technology today. Most of the operational networks are based on WCDMA. It uses new spectrum with 5-MHz carrier. The description in this section is mainly based on UMTS specifications.

5.2   UMTS Network Architecture

The UMTS network is divided into two logical networks, i.e., core network (CN) and radio access network (RAN). The CN and RAN are connected via an open interface. The network architecture is continuously evolving to enable a smooth transition from 2G to 3G, leveraging existing and new technologies. Several specifications based on this evolution have been released by 3GPP.

- 3GPP Release 99/Release 3: Adds 3G radios i.e. UTRAN in enhanced GSM/GPRS core. This provides broadband interface.

- 3GPP Release 4: Adds a softswitch/media gateway in the circuit-switched domain.

- 3GPP Release 5: GERAN, first IP multimedia service (IMS) with SIP, QoS, and IPv6.

- 3GPP Release 6: All IP network, multicast/broadcast multimedia services, WCDMA/WLAN interworking.

5.2.1 3GPP Release 99

Figure 5-1 shows the UMTS architecture as specified in 3GPP Release 99. The system architecture is based on the enhanced GSM Phase 2+ core network with GPRS and a new radio network called UMTS terrestrial radio access network (UTRAN). UTRAN is connected with the core network by the Iu interface.

UTRAN consists of several radio network subsystems (RNSs). An RNS is supported by the core network. Each RNS consists of base stations, termed as Node B in UMTS, and a radio network controller (RNC). The RNC is a BSC equivalent and controls several Node Bs. As shown in Figure 5-1, the 3G terminals (UE) interface with UTRAN using the Uu

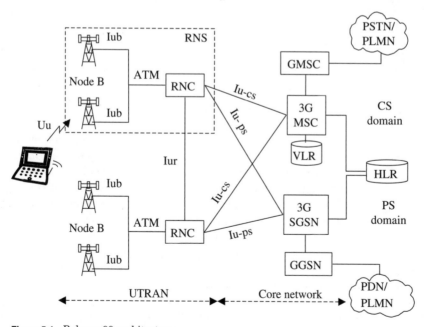

Figure 5-1 Release 99 architecture.

interface, which is a WCDMA-based radio link. The Node Bs are connected to the RNC by Iub interfaces. Unlike the Abis interface, the Iub interface is well defined. This ensures interoperability in a multivendor environment where Node Bs and RNCs are supplied by different vendors. Another point to note here is that, unlike GSM BSCs, Node Bs are connected to each other by the Iur interface. This is required for inter-RNC handover.

A UE may attach to the several RNCs. The RNC that controls Node B is known as controlling RNC (CRNC). It is responsible for managing radio resources for all the Node Bs under its control. The RNC that controls the connection between a UE and the core network is known as a serving RNC (SRNC). In many cases, the CRNC and the SRNC are same. UTRAN supports soft handover. The soft handover occurs between Node Bs supported by different RNCs. During soft handover, the UE starts communicating with the new RNC, i.e., a drift RNC (DRNC), before it takes over the role of SRNC.

As shown in Figure 5-1, the core network consists of network elements to support subscriber control and circuit and packet switching. The core network also supports interfaces to the external network. The RNCs are connected to a 3G MSC by the Iu-CS interface, which supports circuit-switched services. Iu-CS is equivalent to the A interface in GSM. The RNCs are also connected to a 3G SGSN by the Iu-PS interface, which supports packet-switched data services. Iu-PS is equivalent to the Gb interface in GPRS. All the new interfaces, i.e., Iub, Iur, Iu-CS, and Iu-PS, are based on ATM.

In UMTS, the user equipment (UE) or mobile station (MS) comprises mobile equipment (ME) and a UMTS subscriber identity module (USIM).

5.2.2 Release 4 architecture

Figure 5-2 illustrates the Release 4 architecture. As can be noticed, the core network is evolved further and introduces changes in the CS domain. The 3G MSC functions are divided into two parts, i.e., MSC server and media gateways. The MSC server contains call control and mobility management logic. The MSC server also contains a VLR to hold mobile subscriber service data. The media gateway contains the switching function and is controlled by the MSC server. MGW terminates the bearer channels from the circuit-switched network. The same applies to the GMSC server, which is split into GMSC server and media gateway.

Separating the call control and physical interfaces has distinct advantages. It offers scalability and lower cost. Moreover, the information transfer between MS server, media gateways and other components are IP based. Therefore, many components in the core network, including SGSN, GGSN, and MSC server, can be hooked up on the intra PLMN IP

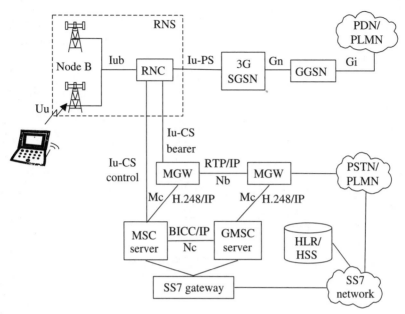

Figure 5-2 Release 4 architecture.

backbone, taking advantage of shared and cheaper IP transport. The MSC server uses ITU-T H.248 to control the media gateway. The ITU-T BICC (bearer-independent call control) protocol is used between the MSC and the GMSC server The core network supports coexistence of both UTRAN and GSM/GPRS radio access network (GERAN).

5.2.3   Release 5 architecture

Figure 5-3 shows the Release 5 architecture. The salient point for this architecture is that it is all IP based. The voice is over IP, and hence there is no need of circuit switching within PLMN. At the gateway, appropriate conversion is required to interconnect to legacy systems. The SGSN and the GGSN are enhanced to support circuit-switched services such as voice. The new roaming signaling gateway (R-SGW) and transport signaling gateway (T-SGW) are needed to provide interworking with the external system over legacy SS7 and SS7-over-IP. The call state control function (CSCF) provides call control functions for multimedia sessions. The media gateway control function (MGCF) controls media gateways, which are IP multimedia subsystems. The media resource function (MRF) supports features such as multiparty conferencing and "meet me."

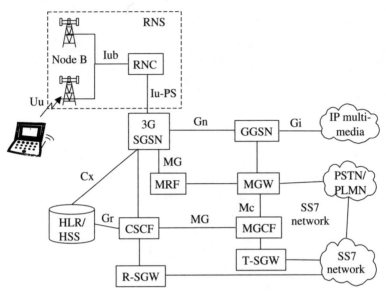

Figure 5-3 Release 5 architecture.

5.3 UMTS Interfaces and Protocols

UMTS leveraged several industry-standard, established protocols. This includes CC, MM, SM (GSM), GTP (GPRS), BICC, SS7, SS7-over-IP/ATM, UDP, IP, and others. However, new protocols have been developed for the UTRAN interfaces. Section 5.4.1 introduces protocol structure at the new Iu interfaces. The detailed specifications for each of these protocols are available from the 3GPP website. Section 5.3.1 describes these protocols in brief.

5.3.1 UTRAN interfaces and protocol structure

Figure 5-4 shows the general protocol model for UTRAN interfaces, i.e. Iub, Iur, Iu-CS, and Iu-PS. The structure consists of two horizontal layers: the Radio Network Layer and the Transport Network Layer.

The Radio Network Layer is concerned with user data and control information. The Transport Network Layer is concerned with the transport technologies used for the UTRAN interfaces. The two layers are logically independent of each other. This makes it possible to change the Transport Network Layer without affecting Radio Network Layer, if required. In Release 99, the Transport Network Layer is based on ATM. In Release 5, IP is used.

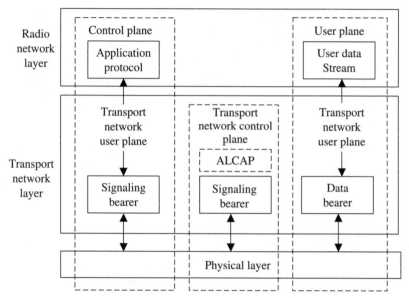

Figure 5-4 Generic protocol model for UTRAN interfaces.

The user plane includes the user data between the UE and the network and the data bearers. The user data consists of data streams characterized by frame protocols specific to a UTRAN interface.

The control plane includes the application protocols and the signaling bearers, which transport the control information. The application protocols used at different UTRAN interfaces are:

- Iu-CS: Radio access network application protocol (RANAP)

- Iu-PS: RANAP

- Iub: Node B application protocol (NBAP)

- Iur: Radio network system application protocol (RNSAP)

The transport network control plane includes the access link control application protocol (ALCAP). ALCAP is used to set up transport bearers to carry user and control plane information. It is not visible to the Radio Network Layer.

Several alternatives are available for the Physical Layer implementation within UTRAN. The specified options in 3GPP release at Iu interfaces are:

- Layer 1 synchronized option, i.e., PDH/SDH/SONET.

- Layer 1 IP nonsynchronized option, i.e., Ethernet or any other suitable point-to-point or point-to-multipoint technique.

Iu-CS interface protocol structure.   In UMTS, the interface between RAN and CN is Iu. Iu-CS is the interface specified between the RAN and the 3G MSC. The Iu-PS interface is defined between the RAN and the 3G SGSN. In order to have uniformity, 3GPP specifies a single protocol at Radio Network Layer for the Iu-CS and the Iu-PS interfaces. The radio access network application protocol (RANAP) is the Radio Network Layer protocol for the Iu interface. The RANAP peer entities reside in 3G MSC/SGSN and the SRNC. The RANAP functions are specified in 3GPP TS 25.413 in detail. In summary, RANAP procedures support the following key functions.

- Radio access bearer (RAB) management including RAB setup, modification, and release

- Iu connection management

- Facilitate general UTRAN procedures from the core network, e.g., paging requests from the CN to UE

- Services to upper layers including the transportation of upper layer nonstratum protocols (i.e., call control, session management, and mobility management) messages between the UE and CN

- Overload and error handling

- SRNS relocation

- UE location reporting

- Trace invocation for a specified UE

- Security functions including ciphering and integrity checks

RANAP uses services provided by the Transport Network Layer to transfer RANAP messages across the Iu interfaces. Figure 5-5 shows the Transport Network Layer protocol stack. The transport layer ensures error free message transfer between two RANAP entities. The Service Connection and Control Part (SCCP) offers both connectionless and connection-oriented services. Each active UE is assigned a separate logical link in case of connection-oriented service between two RANAP entities. The SCCP utilizes services provided by the lower layers to transport messages between two entities. Layer 3 Broadband Message Transfer Part (MTP3b) provides message routing, discrimination, and distribution. It also provides link management functions including load sharing between linksets. The SSCF maps the requirements of above layers to the requirements of SSCOP. The SSCOP provides the mechanism for the establishment and release of connections and the reliable exchange of signaling information between the signaling entities. In cases where the IP transport option is chosen, the services are provided by M3UA, SCTP, and IP. AAL5 is used to adapt the upper layer protocol to the requirements of the lower ATM cells.

The radio network layer–Iu user plane protocols carry the user data over the bearers that are set up by the Transport Network Layer.

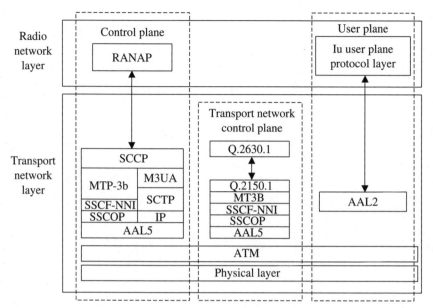

Figure 5-5   Iu-CS interface protocol structure.

The frame protocol in the Iu interface supports both CS and PS domain user data traffic.

As described in the previous section, the purpose of the transport network control plane is to set up, maintain, and release bearers to transport the data via the user plane. The AAL2 signaling protocol capability set 1 (ALCAP), which is described in ITU-T specification Q.2630.1, is used. ALCAP is a Layer 3 protocol. Its responsibility is to set up and manage ATM Adaptation Layer 2 (AAL2) connections.

In the user plane, ATM Adaptation Layer 2 (AAL2) is used as the user data bearer. AAL2 has been specifically designed to transport short-length packets.

Iu-PS interface protocol structure.   The Iu-PS interface is specified between the RAN and the 3G SGSN (Figure 5-6). As described in the previous section, 3GPP specifies a single protocol at the Radio Network Layer for the Iu-CS and the Iu-PS interfaces, i.e., RANAP for the control plane and Iu for the user plane. Both of these are defined in the previous section.

No transport network control protocol is needed. Unlike GPRS, where the GTP tunnel ends at the SGSN, the GTP tunnel in UMTS extends up to RNC. The tunnel ID and IP address, which is required to establish a tunnel, is included in the upper layer protocols.

Like GPRS, GTP-U uses UDP/IP. AAL5 is used to carry the packet-switched user traffic over the Iu-PS interface.

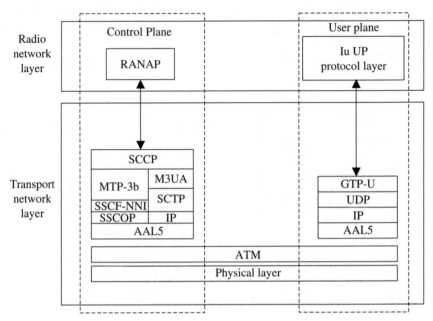

Figure 5-6 Iu-PS interface protocol structure.

Iur interface protocol structure. Iur is the interface between the RNCs (Figure 5-7). One of the RNCs assumes the controlling role and is termed the serving RNC (SRNC); the other RNC is termed the drifting RNC (DRNC).

The Radio Subsystem Application Part (RNSAP) is a Radio Network Layer protocol used at the Iur interface. RNSAP includes procedures for network control signaling between two RNC nodes:

- Radio link management and reconfiguration

- Radio link supervision

- Common control channel (CCCH) signaling transfer

- Paging

- Relocation execution

RNSAP uses the services of the Transport Layer for reliable transfer of signaling messages in both connectionless and connection-oriented modes. The SCCP allows a separate independent logical connection with individual UE. If the ATM transport option is chosen between two RNCs, the SCCP uses MTP3-B, SSCF-NNI, and SSCOP services for networking and routing of messages. In cases where the IP transport option is chosen, these services are provided by the M3UA, SCTP, and IP.

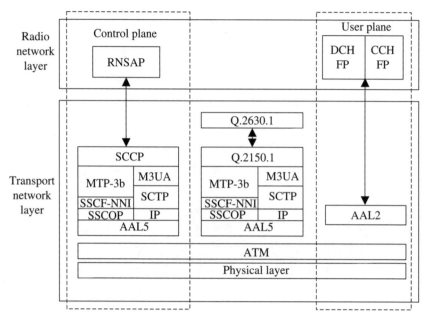

Figure 5-7 Iur interface protocol structure.

Iub interface protocol structure. Iub is the interface between the Node B and the RNC (Figure 5-8).

The Node B application protocol (NBAP) is a Radio Network Layer control plane protocol at the Iub interface. NBAP includes the procedures to manage the logical resources at Node B. NBAP procedures support the following functions:

- Cell configuration management
- Radio link management and supervision
- Common transport channel management
- System information management
- Configuration verification/alignment
- Measurement of common and dedicated resources

5.3.2   System network protocols

Access and nonaccess stratum protocols. The Access Stratum (AS) is defined as the group of protocols (all layers) embedded in the UTRAN and between the edge nodes (UE and RNC). The Nonaccess Stratum (NAS) is defined as the group of protocols between the UE and the CN. These protocols are carried transparently through the UTRAN. Figures 5-9 and 5-10 show the Access Stratum and Nonaccess Stratum protocol boundaries in the control and user planes, respectively.

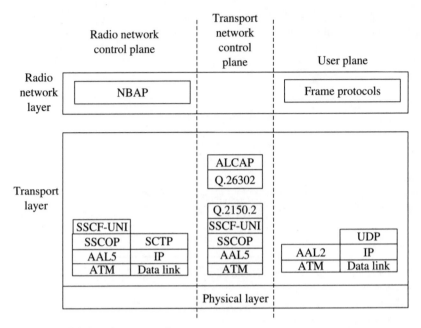

Figure 5-8   Iub interface protocol structure.

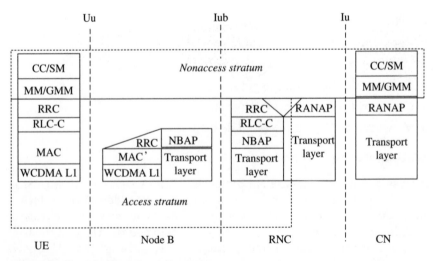

Figure 5-9   Control plane protocols.

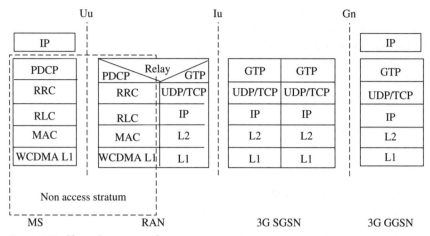

Figure 5-10 User plane protocols.

RAN protocols (Access Stratum) are described in the previous section. The protocols belonging to the Nonaccess stratum group are:

- CS domain
    - Call control management (CC)
    - Mobility management (MM)
- PS domain
    - Session management (SM)
    - GPRS mobility management (GMM)

The call control management protocol contains the functions and procedures for call establishment, monitoring, and release for circuit-switched voice and multimedia calls. The UMTS CC protocol is an extension of the GSM CC protocol described in Chapter 3. The mobility management protocol includes procedures for UE mobility and authentication. As shown in Figure 5-9, the CC protocol uses the connection service provided by the MM sublayer. The MM sublayer in turn uses the connection services provided by the Radio Resource Connection (RRC) Layer. The RRC handles the control plane signaling of Layer 3 between the UE and UTRAN. It establishes, maintains, and releases the signaling connection and radio bearers for UE on request from the upper layer. The RNC uses the relay functionality to map the CC messages into RANAP for forward transmission to core network.

Like the CC protocol, the session management protocol is used to activate, modify, and delete PDP contexts. The prerequisite to a SM context for a UE is the existence of a GMM context. The GMM includes the functions and procedures for the UE mobility and authentication procedure in the PS domain. GMM and SM are the same protocols used in GPRS and are described in Chapter 4.

5.4 Example UMTS Procedures

5.4.1 Mobile-originated circuit-switched calls

The steps to establish an MOC are as follows:

Step 1: RRC connection setup between UE and SRNC

Step 2: Authentication and ciphering

Step 3: Radio access bearer establishment and call setup

Step 4: Call and Iu release

Step 1: RRC connection setup between UE and SRNC. Figure 5-11 illustrates the interaction within UTRAN to establish an RRC connection between the UE and the RNC. The process to set up a call begins with the UE sending an **RRC connection request** over a CCCH (which is a RACH in the uplink direction). This message contains several

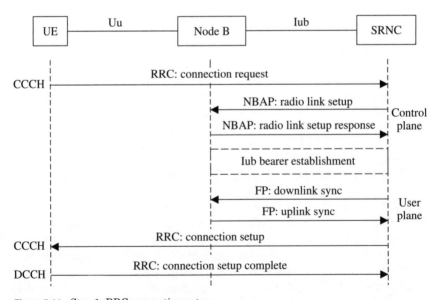

Figure 5-11 Step 1: RRC connection setup.

information elements, including IMSI or TMSI, LAI, RAI, and the reason for requesting the RRC connection.

The RNC analyzes the reason for the request in order to decide the appropriate resources, i.e., dedicated or common. The RNC then initiates the process to establish an Iub bearer by sending the NBAP **radio link setup** message to Node B. This message contains information elements such as the transaction ID, communication ID, scrambling code, transport format set, and FDD-DL channelization code number. The Node-B acknowledges this message by sending an NBAP **RL setup response**. This message contains the information related to Transport Layer addressing information, i.e., AAL2 address. The SRNC uses ALCAP in the Transport Network Layer to establish an Iub bearer, using the information received from the Node B, i.e., AAL path and channel ID. The Iub bearer is bound together with the DCH assigned to the transaction. The SRNC then synchronizes the frame protocol (FP) connection by sending an FP **downlink sync message**. The RNC responds to the UE, indicating a successful RRC connection by sending an RRC **connection setup message**. This message contains information elements such as transport format, power control, and scrambling code. The UE responds with the RRC **connection setup complete** to confirm the RRC connection establishment.

Step 2: Authentication and ciphering.   On successful connection setup with the RNC, the UE sends the RRC **initial direct transfer** message. This message is destined to the core network. However, the RNC processes this partially, adds some more information needed to set up a call and map it to the RANAP **UE initial message**. and sends it to the 3G MSC. The information elements within this message carry information on UE identity, location, and connection setup requirements. This message also indicates to the MSC and the RNC that a new signaling relationship between the UE and CN needs to be established.

On receiving the service request from the UE, the MSC initiates the security procedures. This includes the UE authentication and exchange of the encryption key. The MSC sends an **authentication request** within the RANAP **direct transfer** message. The RNC maps and forwards the **authentication request** message using RRC **direct transfer** to UE. The UE executes the authentication algorithm and sends the result back in an **authentication response** message to the MSC. As shown in Figure 5-12, this message is carried over as payload in the RRC **direct transfer** and RANAP **direct transfer** messages. The RNC merely acts as a relay. Assuming that the UE is successfully authenticated, the MSC then sends a **security mode command** to the RNC indicating that the further transactions between the UE and the UTRAN should be encrypted. The RNC in turn sends an RRC **security mode command** message to UE. The security mode command message conveys the encryption algorithm and

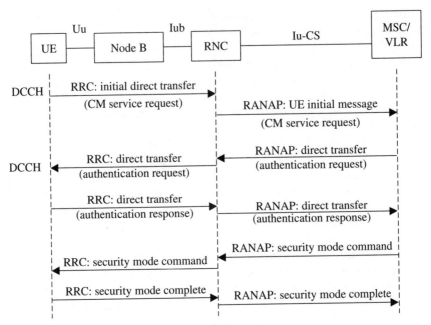

Figure 5-12 Step 2: Authentication and ciphering.

the encryption and integrity keys. The UE starts encrypting any further transaction toward UTRAN and informs the RNC, using a **RRC security mode complete** message. The RNC in turn informs the MSC. Note that encryption is applied only on the transaction between the UTRAN and the UE.

Step 3: Radio access bearer establishment and call setup. After the successful authentication and security procedures, the UE sends a call control **setup** message to the MSC. The MSC verifies that the UE is authorized for the requested services. If yes, the MSC starts a process to set up a bearer for the user data (speech in this case). This is achieved by the MSC by sending an RAB assignment request to the RNC (Figure 5-13). The MSC includes the RAB ID and the QoS parameters to be set up. The RNC, on receiving this message, checks the resources and sets up a bearer at Iu. The actual bearers are set up by using the ALCAP in the Network Transport Layer. The ALCAP procedures are not shown in the figure. The RNC in turn sets up a radio bearer between the RNC and the UE by sending a **radio bearer setup** message. This message contains the information on bearer allocation, i.e., a radio bearer identifier. The UE responds with the **radio bearer setup complete** message. The RNC then sends

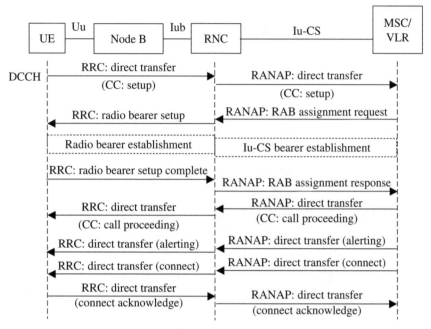

Figure 5-13 Step 3: RAB establishment and call setup.

an **RAB assignment response** to the MSC. With this procedure successfully executed, there exists a bearer to transport used data from the UE to the MSC.

From this point onward, the call proceeds in a normal way, using call control messages as in GSM call setup.

Step 4: Call and RAB release. Once the call is released by any of the parties, the resources need to be released. As shown in Figure 5-14, on receiving a disconnect message from the UE (in this example, the calling party releases the call) and transfer of subsequent call clearing messages, the MSC issues an **Iu release command** to the RNC. On receiving this message, the RNC releases the radio bearer over Iub interface and informs the MSC by sending an **Iu release complete message**. Now the RNC takes charge to clear the RRC connection by sending an RRC **connection release** message to the UE. The UE acknowledges with a **connection release complete** message.

The last action for the RNC is to clear the Iub interface resources. The procedure is illustrated in Figure 5-15. The MSC sends an NBAP **radio link deletion** message to the Node B. The Node B responds with a

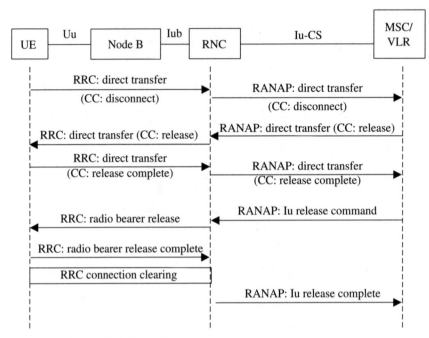

Figure 5-14   Step 4(a): Call clearing.

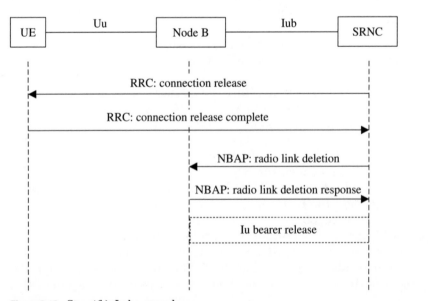

Figure 5-15   Step 4(b): Iu bearer release.

radio link deletion response message to indicate the release of Iub interface resources.

5.4.2 Mobile-originated packet-switched calls

In general, the steps defined in the previous section to establish a circuit-switched call are also followed to establish a packet-switched call. However, as one can understand, the procedures used are somewhat different.

Step 1: RRC connection setup between UE and SRNC. The same procedures are followed as in the case of a circuit-switched call except that the reason indicated in the RRC connection request message is a data call.

Step 2: Authentication and ciphering. The same procedures are followed as in the case of circuit-switched call except that the authentication and security procedures are invoked with the serving SGSN, as shown in Figure 5-16

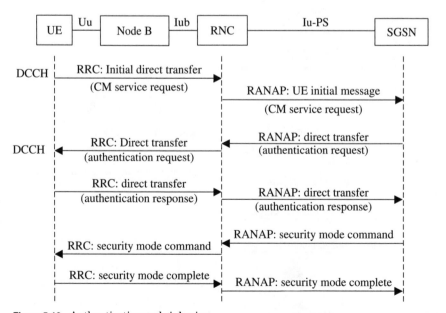

Figure 5-16 Authentication and ciphering.

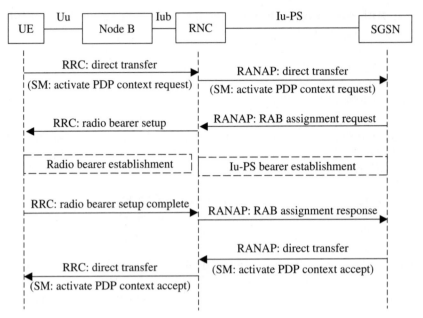

Figure 5-17 RAB and PDP context establishment.

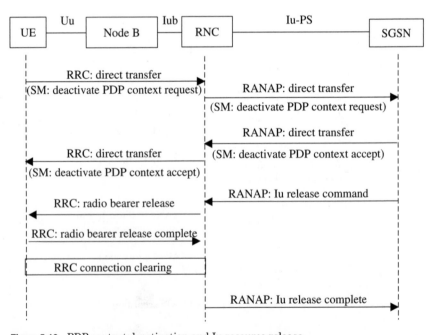

Figure 5-18 PDP context deactivation and Iu resource release.

Step 3: Radio access bearer establishment and PDP context activation. As shown in Figure 5-17, the main difference in a packet-switched call is that the session management (SM) protocol is used instead of the call control protocol.

Step 4: PDP context deactivation and Iu release. Figure 5-18 shows the Iu release procedure when the UE deactivates the PDP context.

Bibliography

3GPP TS 23.002, Network Architecture.
3GPP TS 25.401, UTRAN Overall description.
3GPP TS 25.413, UTRAN Iu Interface RANAP Signalling.
3GPP TS 25.423, UTRAN Iur Interface RNSAP Signalling
3GPP TS 25.433, UTRAN Iub Interface NBAP Signalling.
3GPP TS 24.008, Mobile radio interface Layer 3 Specification.

Chapter

6

Roaming in a GSM Network

6.1 Inter-PLMN Signaling Network

A prerequisite for international roaming is connectivity between a Home Public Land Mobile Network (HPLMN) and a Visited Public Mobile Network (VPLMN) for signaling and bearer services, e.g., voice and data.

Figure 6-1 shows that the HPLMN is connected with the VPLMN via the international public switched telephone network (PSTN) for bearer services. This consists of 64-Kbps circuit-switched voice or data links. The signaling required for ISUP calls and also to enable roaming is carried over a logically separate network.

The signaling network carries MAP messages, using SCCP and MTP. An HPLMN and a VPLMN are connected either directly or via an international signaling network. GSM operators normally use an international hub to avoid more expensive point-to-point CCS7 links. However, GSM operators also connect directly to the partner networks that carry heavy roaming traffic, e.g., neighboring countries. GSM operators usually partner with more than one operator in a foreign country to ensure reliability.

The international signaling network consists of SCCP gateways and STPs. It transports MAP signaling messages between PLMNs. An end-to-end partnership agreement must exist between cooperating PLMNs.

Figure 6-2 shows a PLMN in a home country connected to PLMN 1 and 2 of country X and PLMN 2 of country Y. There is no roaming agreement between the PLMN in the home country and the PLMN 1 of country Y.

For international roaming, network nodes in a VPLMN need to communicate with those of a subscriber's HPLMN. For example, the visited network needs to verify if a foreign subscriber trying to register in its network is authorized and has subscribed to the roaming services.

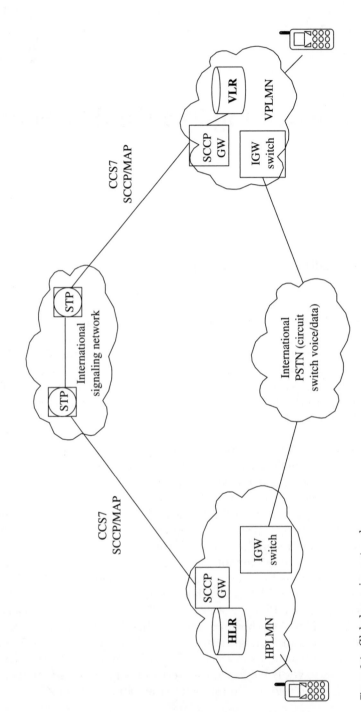

Figure 6-1 Global roaming network.

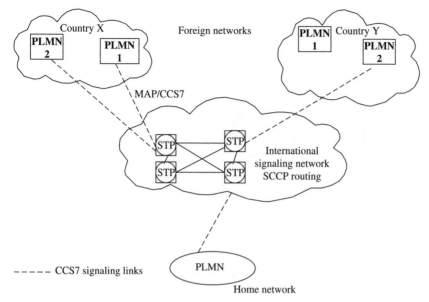

Figure 6-2   Inter-PLMN signaling network.

6.1.1   Inter-PLMN addressing

The MAP protocol uses SCCP addressing capabilities to route signaling messages between the VPLMN and the HPLMN nodes across the international network.

The SCCP calling and called party addresses contain the necessary information for the SCCP to route messages between PLMNs. The SCCP message routing is performed using global title (GT) or destination point code (DPC) and subsystem number (SSN). The format and coding of the calling and the called address parameters comply with the SCCP addressing guidelines as defined in ITU-T Q.713 Recommendations. Section 6.1.2 provides a detailed description of the SCCP addressing capabilities.

6.1.2   Address format

The SCCP calling and called party address format for inter-PLMN message routing is illustrated in Figure 6-3.

SCCP called party address (CgPA).   The called party address is a variable-length parameter. Its structure is as follows:

- Point code indicator (PCI) = 0 indicates that the address does not contain a signaling point code.
- SSN indicator (SSNI) = 1 indicates that MAP SSN is included.

8	7	6	5	4	3	2	1	
0	RI = 0	GTI = 4				SSNI = 1	PCI = 0	Octet 1
SSN = 0 or international standard value								Octet 2
Translation type = 0								Octet 3
Numbering plan = 1 (E.164)				Encoding scheme = 1 or 2				Octet 4
0				Nature of address indicator = 4 (international)				Octet 5
Country code digit 2 (if present)				Country code digit 1				Octet 6
National destination code (NDC) digit 1				Country code digit 3 (if present)				Octet 7
NDC digit 3 (if present)				NDC digit 2 (if present)				Octet 8
NDC digit 5 (if present)				NDC digit 4 (if present)				Octet 9
Equipment identification digit 2				Equipment identification digit 1				Octet 10
·				·				·
If needed, filler = 0				Equipment identification digit N (if present)				Octet M

Figure 6-3   Inter-PLMN addressing format.

- Global title indicator (GTI) = 0100 indicates that the global title includes translation type, numbering plan, encoding scheme, and nature of address indicator. This GTI coding (i.e., 0100) is used for international network applications.

- Routing indicator = 0 indicates routing is based on global title.

- The last bit of the octet is reserved for national use and is always set to zero for the international network.

- Subsystem number (SSN) = 0 or international standard values as given in Table 6-1.

- For translation type = 0, numbering plan =1, and nature of address indicator = 4, the address is coded according to the following rules for international GT routing.

  - A maximum of the first three digits are used to identify the destination country or region of the addressable entities. For destination countries with only one operator, translation of the country code (CC) is sufficient.

  - For destination countries with multiple network operators, the CC and network destination code (NDC) are translated within the international network to identify the destination.

  - Translation of additional digits (i.e., equipment identification) to identify a specific SCCP user entity is a national matter or network specific.

SCCP calling party address(CdPA).   The calling party address is a variable-length parameter. Its structure is the same as the called party address.

Figure 6-4 shows a protocol decode for SCCP message routing based on GT. In this example, the VLR in a Hong Kong network is trying to

TABLE 6-1   SSN Identifiers

SSN	Subsystem
0	SSN not known
1	SCCP management
2	Reserved
3	ISUP
4	OMAP
5	MAP
6	HLR
7	VLR
8	MSC
9	EIR
10	AUC
11	SMSC

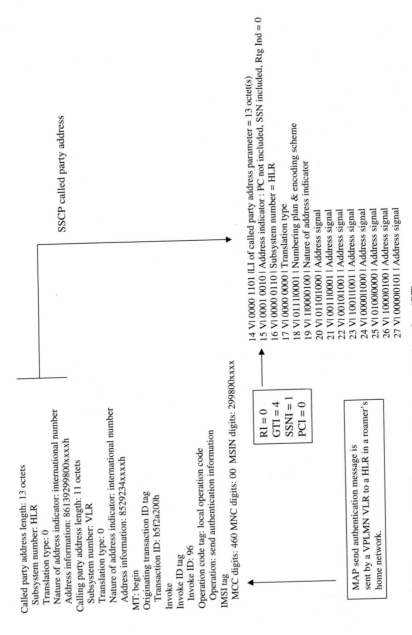

SSCP called party address

Called party address length: 13 octets
Subsystem number: HLR
Translation type: 0
Nature of address indicator: international number
Address information: 8613929980xxxh
Calling party address length: 11 octets
Subsystem number: VLR
Translation type: 0
Nature of address indicator: international number
Address information: 8529234xxxh
MT: begin
Originating transaction ID tag
Transaction ID: b5f2a200h
Invoke
Invoke ID tag
Invoke ID: 96
Operation code tag: local operation code
Operation: send authentication information
IMSI tag
MCC digits: 460 MNC digits: 00 MSIN digits: 299800xxxx

RI = 0
GTI = 4
SSNI = 1
PCI = 0

14 V| 0000 1101 ILI of called party address parameter = 13 octet(s)
15 V| 0001 0010 | Address indicator : PC not included, SSN included, Rtg Ind = 0
16 V| 0000 0110 | Subsystem number = HLR
17 V| 0000 0000 | Translation type
18 V| 0111|0001 | Numbering plan & encoding scheme
19 V| 1|0001|000 | Nature of address indicator
20 V| 0|110|1000 | Address signal
21 V| 001|1|0001 | Address signal
22 V| 0010|1001 | Address signal
23 V| 1001|1001 | Address signal
24 V| 0000|1000 | Address signal
25 V| 0100|0000 | Address signal
26 V| 1000|0100 | Address signal
27 V| 0000|0101 | Address signal

MAP send authentication message is
sent by a VPLMN VLR to a HLR in a roamer's
home network.

Figure 6-4 Example MAP message decode -routing based on 'GT'

142

authenticate a roamer (subscriber from China roaming in Hong Kong) by sending a send authentication message to the subscriber's PLMN HLR. The routing of this message is based on global title.

6.2 Communication Between a VPLMN VLR and an HPLMN HLR

When a roamer switches ON a mobile station (MS) for the first time in a VPLMN, the VLR initiates the update location procedure with the roamer's HLR. The only information available to the VPLMN VLR at this time is the IMSI of the roamer. The VPLMN VLR uses this to derive routing information (SCCP addressing) for communicating with the HPLMN HLR. The derived address is known as the mobile global title (MGT) or E.214 address.

When responding to the VPLMN VLR, the HPLMN HLR inserts its own E.164 address in the CgPA of the SCCP message. The E.164 part, as defined in the ITU-T E.164 Recommendation, is used to identify the country and PLMN or PLMN and HLR, where the roamer is registered.

On receiving an initial response from the HPLMN HLR, a VPLMN VLR then derives the routing information for subsequent communication with the HPLMN HLR from the calling party address in the received response.

This means that the VPLMN VLR is able to address the HPLMN HLR using an E.214 MGT that has been originally derived from the roamer's IMSI and an E.164 HLR address.

An E.214 MGT translation is done either at the application level or at the SCCP level in the VLR using a routing table.

As shown in Figure 6.5., the MGT is of variable length. Within the MGT, the Country Code (CC) is derived from the Mobile Country Code (MCC). The National Destination Code (NDC) is derived from Mobile Network Code (MNC) or from the MNC and some initial digits of the Mobile Station Identifier Number (MSIN).

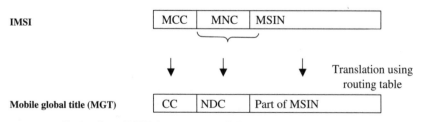

Figure 6-5 Derivation of MGT from roamer's IMSI.

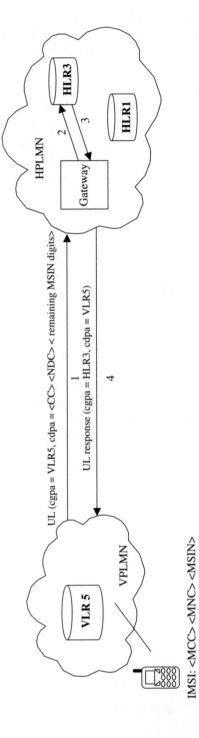

IMSI: <MCC> <MNC> <MSIN>

Figure 6-6 MGT-based routing.

Each PLMN consists of one logical HLR. In practical implementations, one physical HLR covering an entire network may not be feasible. In most of the implementations, more than one HLR may exist, grouped under one logical HLR. The SCCP gateway/GMSC at the edge of a network decides to route the message received by a VPLMN to the right HLR on the basis of the MGT in the SCCP called party address. As shown in Figure 6-6, VLR5 derives SCCP called party address (CdPA) from the roamer's IMSI. The gateway MSC in the HPLMN makes the decision to route the message to HLR 3 by looking at NDC and MSIN digits.

Figure 6-7 shows partial protocol decodes of an update location request and an update location response message. In this example, a roamer from a Singapore network is trying to register in a network in Malaysia. The serving VLR uses MGT translation to route the UL message to the HLR in the roamer's home PLMN in Singapore. The HPLMN GMSC routes the UL request to the HLR, which contains subscriber information. The HLR, in a response message, inserts its own address in the CgPA. The CgPA received in the request message is used as the CdPA in subsequent messages from the VLR.

6.3   Roaming Procedures

This section describes the information transfer and procedures that take place between the VPLMN and the HPLMN to enable roaming. On first-time roamer registration in a VPLMN, the serving VLR updates the HPLMN HLR with the new location of the MS (i.e., the VLR address), using the update location procedure. The HLR uses the cancel location procedure with the first VPLMN when a roamer appears in a different PLMN or returns to the HPLMN. The VPLMN VLR invokes the subscriber parameter request procedure (at any time) to request the HLR to provide subscriber parameters for a specified roamer. A VLR may use the purge procedure to inform the HLR of the deletion of data for a roamer that has not established radio contact for a specified period. On restart of the HLR after a failure, recovery and restore procedures are invoked by the HLR.

6.3.1   Location update in a visited network

The update location procedure is initiated by the VPLMN VLR when –

- A roamer turns ON an MS for the first time in a foreign network, i.e., to attach/register.

- A roamer moves to a new location area (LA), i.e., forced registration.

- The VLR receives instructions as a consequence of an HLR or a VLR restoration.

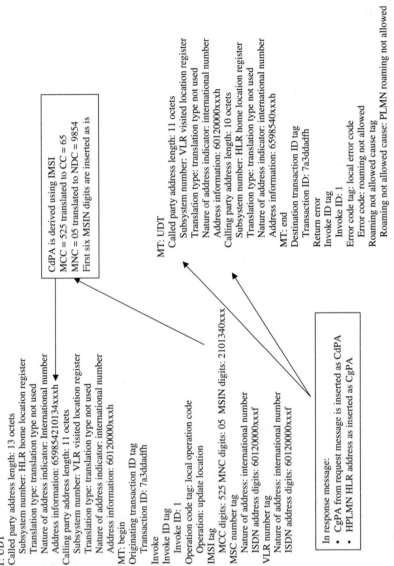

MT: UDT
  Called party address length: 13 octets
    Subsystem number: HLR home location register
    Translation type: translation type not used
    Nature of address indicator: International number
    Address information: 6598542101340xxxh
  Calling party address length: 11 octets
    Subsystem number: VLR visited location register
    Translation type: translation type not used
    Nature of address indicator: international number
    Address information: 60120000xxxh
MT: begin
  Originating transaction ID tag
    Transaction ID: 7a3ddadfh
Invoke
  Invoke ID tag
    Invoke ID: 1
  Operation code tag: local operation code
    Operation: update location
  IMSI tag
    MCC digits: 525 MNC digits: 05 MSIN digits: 2101340xxx
  MSC number tag
    Nature of address: international number
    ISDN address digits: 60120000xxxf
  VLR number tag
    Nature of address: international number
    ISDN address digits: 60120000xxxf

CdPA is derived using IMSI
MCC = 525 translated to CC = 65
MNC = 05 translated to NDC = 9854
First six MSIN digits are inserted as is

MT: UDT
  Called party address length: 11 octets
    Subsystem number: VLR visited location register
    Translation type: translation type not used
    Nature of address indicator: international number
    Address information: 60120000xxxh
  Calling party address length: 10 octets
    Subsystem number: HLR home location register
    Translation type: translation type not used
    Nature of address indicator: international number
    Address information: 6598540xxxh
MT: end
  Destination transaction ID tag
    Transaction ID: 7a3ddadfh
Return error
  Invoke ID tag
    Invoke ID: 1
  Error code tag: local error code
    Error code: roaming not allowed
  Roaming not allowed cause tag
    Roaming not allowed cause: PLMN roaming not allowed

In response message:
• CgPA from request message is inserted as CdPA
• HPLMN HLR address as inserted as CgPA

Figure 6-7 MGT translation.

The VLR also initiates the update location procedure periodically to ensure that mobiles are not detached accidentally from the system.

The map update procedure is used by the VPLMN VLR to update the location information stored in the HPLMN HLR.

Figure 6-8 illustrates the location update procedure.

The following information fields/parameters are sent from the serving MSC/VLR in the visited network to the HLR in the roamer's home country.

- IMSI
- MSC address
- VLR number

The HPLMN HLR uses IMSI as the key to extract the roamer's information from its database. The MSC address is the E.164 address of the serving MSC. The HLR stores the serving MSC address (i.e., the current location of the roamer) in its database. The VLR number is the E.164 ISDN number of the VLR. The HPLMN HLR uses the MSC address and the VLR number in subsequent roaming procedures and call handling.

Figure 6-9 shows the protocol decodes for a MAP update location request message invoked by a serving VLR in a visited network.

As part of the location update procedure, the HPLMN HLR sends the subscriber parameters of the roamer to the VPLMN VLR. The HPLMN HLR

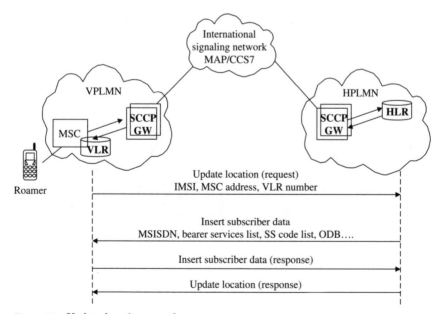

Figure 6-8 Update location procedure.

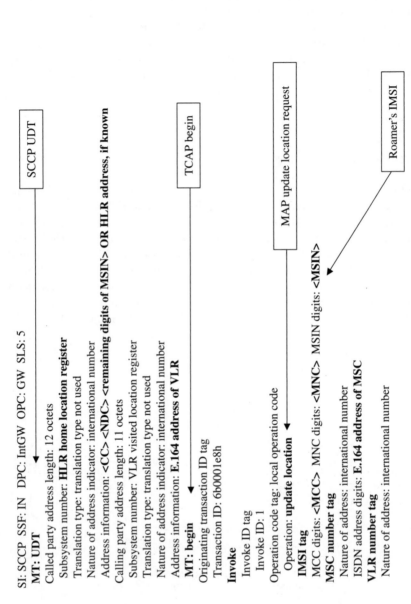

SI: SCCP SSF: IN   DPC: IntGW OPC: GW SLS: 5

SCCP UDT

MT: UDT

Called party address length: 12 octets
Subsystem number: **HLR home location register**
Translation type: translation type not used
Nature of address indicator: international number
Address information: **<CC> <NDC> <remaining digits of MSIN> OR HLR address, if known**

Calling party address length: 11 octets
Subsystem number: VLR visited location register
Translation type: translation type not used
Nature of address indicator: international number
Address information: **E.164 address of VLR**

TCAP begin

MT: begin

Originating transaction ID tag
Transaction ID: 6b0001e8h

Invoke

Invoke ID tag
  Invoke ID: 1

MAP update location request

Operation code tag: local operation code
  Operation: **update location**

IMSI tag

MCC digits: <MCC> MNC digits: <MNC> MSIN digits: **<MSIN>**

Roamer's IMSI

MSC number tag

Nature of address: international number
ISDN address digits: **E.164 address of MSC**

VLR number tag

Nature of address: international number

Figure 6-9   Example message decodes for an update location request.

invokes the MAP insert subscriber data procedure to update the VPLMN VLR. The important parameters sent by the HLR are as follows -

- *MSISDN. The ISDN number assigned to the roamer.*
- *IMSI.* The IMSI of the roamer.
- *Category.* This refers to the calling party category. It is included either at location updating or when it is changed.
- *Subscriber status.* This parameter is set to service granted if no operator-determined barring (ODB) is required. To apply, remove, or update ODB, the subscriber status is set to operator-determined barring. In this case, the ODB parameter is present.
- *Forwarding information list.* This includes the SS code for an individual call forwarding supplementary service.
- *Call barring information list.* This includes the SS code for an individual call barring supplementary service.
- *CUG information list.* This includes the complete subscribed CUG feature list.
- *Bearer service list.* This lists the codes of all bearer services subscribed to by the roamer.
- *Teleservice list.* This lists the codes of all the teleservices subscribed to by the roamer.
- *Operator-determined barring HPLMN data.* This includes all the operator-determined barring categories that may be applied to a roamer in a VPLMN.
- *SS code list.* This lists the SS codes for individually subscribed supplementary services. It is sent for supplementary services other than call forwarding, call barring, and CUG.

Figure 6-10 shows a partial decode of a MAP insert subscriber data operation code.

In the case of unsuccessful updating, the HPLMN HLR responds with an error message. An error cause is included to inform VPLMN VLR of the reason for failure.

In the event of failures, several different error causes can be returned:

- *Unknown subscriber.* No such subscriber.
- *Roaming not allowed.* The diagnostic code provides further information.
  - PLMN not allowed
  - Operator-determined barring

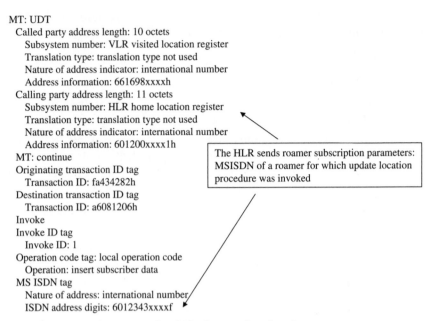

MT: UDT
  Called party address length: 10 octets
    Subsystem number: VLR visited location register
    Translation type: translation type not used
    Nature of address indicator: international number
    Address information: 661698xxxxh
  Calling party address length: 11 octets
    Subsystem number: HLR home location register
    Translation type: translation type not used
    Nature of address indicator: international number
    Address information: 601200xxxx1h
MT: continue
Originating transaction ID tag
  Transaction ID: fa434282h
Destination transaction ID tag
  Transaction ID: a6081206h
Invoke
Invoke ID tag
  Invoke ID: 1
Operation code tag: local operation code
  Operation: insert subscriber data
MS ISDN tag
  Nature of address: international number
  ISDN address digits: 6012343xxxxf

The HLR sends roamer subscription parameters: MSISDN of a roamer for which update location procedure was invoked

Figure 6-10 Example message decode for insert subscriber data.

PLMN not allowed is used as the default if no qualifying information is received.

■ *Unexpected data value.* The data type is formally correct but its value or presence is unexpected in the current context.

■ *System failure.* The task cannot be performed because of a problem in the other node. The type of node or network resource may be indicated by use of the network resource parameter.

■ *Data missing.* An optional parameter required by the context is missing.

Figure 6-11 shows an HPLMN HLR responding with an error stating that this subscriber is not allowed to roam in a foreign network.

6.3.2 Roamer authentication in visited network

To ensure security and to deny services to an unauthorized visitor, the VPLMN has to validate a roamer as an authorized subscriber before granting permission to roam in its network. The VPLMN may also authenticate roamers periodically on observing further activities. Inter-PLMN authentication of a roamer follows the same procedure as defined

SI: SCCP SSF: IN   DPC: GW OPC: IntGW SLS: 13

MT: UDT

Called party address length: 11 octets

Subsystem number: VLR visited location register

Translation type: translation type not used

Nature of address indicator: international number

Address information: **E.164 VLR address received in location update request message**

Calling party address length: 11 octets

Subsystem number: **HLR home location register**

Translation type: translation type not used

Nature of address indicator: international number

Address information: **E.164 HLR address**

MT: end

Destination transaction ID tag

Transaction ID: 6b0001e8h

Return error

Invoke ID tag

Invoke ID: 1

Error code tag: local error code

Error code: **roaming not allowed**

Roaming not allowed cause tag

Roaming not allowed cause: **operator determined barring**

SCCP UDT

TCAP END

MAP update location response in error

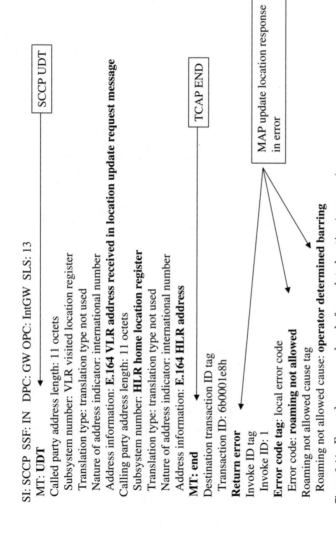

Figure 6-11 Example message decode for update location (response).

151

in Chapter 2 for local subscribers. In this case, however, triplets, i.e., RAND, SRES, and Kc are sent between PLMNs over connecting networks.

The VPLMN VLR initiates the MAP send authentication info procedure to retrieve authentication information from the HPLMN HLR. The response from an HPLMN HLR contains the authentication set list parameter. This list contains a set of authentication triplets, i.e., RAND, SRES, and Kc.

The transfer of authentication key triplets between the HPLMN HLR and the VPLMN VLR is shown in Figure 6-12. The sequence shown is according to the MAPv2 protocol specification. For MAPv1, the MAP send parameter (authentication sets) procedure is invoked instead.

A set of one to five authentication triplets is transferred from the HPLMN HLR to the VPLMN VLR for a successful authentication.

If the VLR receives a MAP send authentication info response containing a user error parameter as part of the handling of an authentication procedure, the authentication procedure in the VLR will fail.

The IMSI of the roamer is sent as a parameter within the MAP send authentication info request message from the VPLMN MSC/VLR to the HPLMN HLR. This is used by the HLR to identify roamer.

The MAP send authentication info response from HPLMN HLR to VPLMN MSC/VLR consists of following parameters.

- Authentication set list

- User error (if present)

Figure 6-13 shows the HLR provided four sets of triplets to the VLR in response to the send authentication request.

In the event of failure, one of the following error causes can be returned:

- *Unknown subscriber.* There is no allocated IMSI or no directory number for the mobile subscriber in the HLR.

- *Unexpected data value.* The data type is formally correct, but its value or presence is unexpected in the current context.

- *System failure.* The task cannot be performed because of a problem in the other node. The type of node or network resource may be indicated in the network resource parameter.

- *Data missing.* An optional parameter that is required by the context is missing.

6.3.3   Provide roaming number

This section describes the roaming procedures initiated by the HPLMN to route terminating calls to its subscribers when roaming

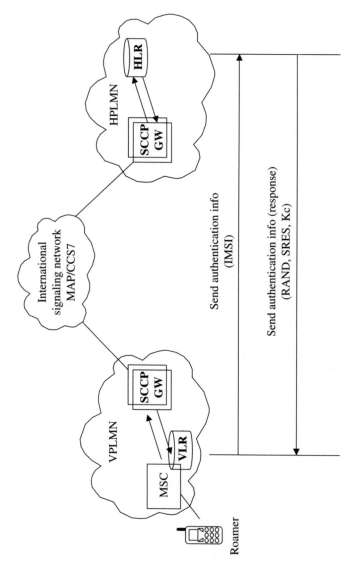

Figure 6-12 Send authentication procedure.

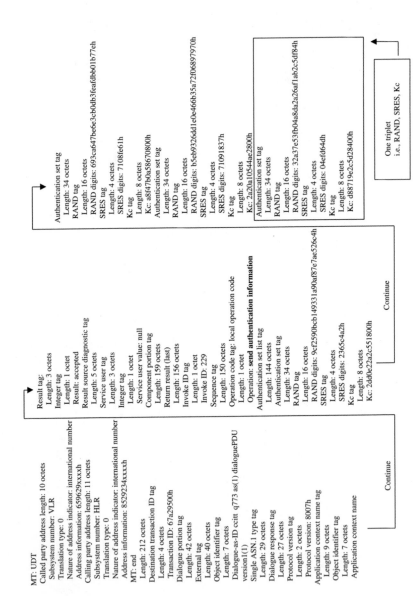

MT: UDT
Called party address length: 10 octets
Subsystem number: VLR
Translation type: 0
Nature of address indicator: international number
Address information: 659629xxxh
Calling party address length: 11 octets
Subsystem number: HLR
Translation type: 0
Nature of address indicator: international number
Address information: 8529234xxxh
MT: end
Length: 212 octets
Destination transaction ID tag
Length: 4 octets
Transaction ID: 67a29500h
Dialogue portion tag
Length: 42 octets
External tag
Length: 40 octets
Object identifier tag
Length: 7 octets
Dialogue-as-ID ccitt q773 as(1) dialoguePDU version1(1)
Single ASN.1 type tag
Length: 29 octets
Dialogue response tag
Length: 27 octets
Protocol version tag
Length: 2 octets
Protocol version: 8007h
Application context name tag
Length: 9 octets
Object identifier tag
Length: 7 octets
Application context name

Continue

Result tag:
Length: 3 octets
Integer tag
Length: 1 octet
Result: accepted
Result source diagnostic tag
Length: 5 octets
Service user tag
Length: 3 octets
Integer tag
Length: 1 octet
Service user value: null
Component portion tag
Length: 159 octets
Return result (last)
Length: 156 octets
Invoke ID tag
Length: 1 octet
Invoke ID: 229
Sequence tag
Length: 150 octets
Operation code tag: local operation code
Length: 1 octet
Operation: **send authentication information**
Authentication set list tag
Length: 144 octets
Authentication set tag
Length: 34 octets
RAND tag
Length: 16 octets
RAND digits: 9cf25900bcb149331a90af87e7ae526c4h
SRES tag
Length: 4 octets
SRES digits: 2365c4a2h
Kc tag
Length: 8 octets
Kc: 2dd0e22a2c551800h

Continue

Authentication set tag
Length: 34 octets
RAND tag
Length: 16 octets
RAND digits: 693ca647be6c3cb0db3feafdbb01b77eh
SRES tag
Length: 4 octets
SRES digits: 7108fe61h
Kc tag
Length: 8 octets
Kc: a8f47b0a58670800h
Authentication set tag
Length: 34 octets
RAND tag
Length: 16 octets
RAND digits: b5eb9326dd1e0e466b35a72f06897970h
SRES tag
Length: 4 octets
SRES digits: 71091837h
Kc tag
Length: 8 octets
Kc: 2a20a105544ae2800h
Authentication set tag
Length: 34 octets
RAND tag
Length: 16 octets
RAND digits: 32a37e53f0b04a8da2a26af1ab2c5df84h
SRES tag
Length: 4 octets
SRES digits: 04efd64dh
Kc tag
Length: 8 octets
Kc: d88719e2c5d28400h

One triplet
i.e., RAND, SRES, Kc

Figure 6-13 Example message decode - Send Authentication Info

in a foreign network. The first step for the HPLMN MSC responsible for routing the call is to get routing information from the local HLR. This is done via the send routing information (SRI) procedure that takes place between the MSC and the HLR. The HLR, on receiving a SRI request, checks the subscriber's status, and, on determining that the subscriber is in a foreign network, invokes the provide roaming number (PRN) procedure toward the VPLMN VLR to get the temporarily assigned mobile subscriber roaming number (MSRN) to the roamer.

Figure 6-14 shows the send routing info and provide roaming number procedures. The detailed steps are as follows.

1. The GMSC responsible for routing a terminating call invokes an MAP send routing information request toward the HLR with the called party MSISDN as the key parameter.

2. The HLR looks at its database and, on finding that the called subscriber is visiting a foreign network, invokes the MAP provide roaming number procedure toward the VPLMN VLR with the roamer's IMSI and MSISDN and the E.164 MSC number currently serving the MS.

3. The serving VPLMN VLR checks if the roamer is still in its network and assigns a temporarily number for routing purposes, i.e., MSRN. If the VPLMN VLR is unable to assign a MSRN to the roaming MS, the provide roaming number procedure fails. The VPLMN VLR then returns a response with the appropriate error code. On successful assignment, the VPLMN VLR sends MSRN in response to PRN request.

4. On receiving a successful response from the VPLMN VLR, the HLR sends an acknowledgment to the GMSC with the MSRN and the VMSC (visited MSC) address.

5. The MSC then invokes normal ISUP procedures toward the VPLMN to route the incoming call to the roamer.

Figure 6-15 shows the protocol decodes for the provide roaming number procedure. In this example, the MSRN is successfully delivered to the HLR.

The errors associated with the provide roaming number procedure are as follows.

- *Absent subscriber.* This indicates that the location of the roamer is not known (either the roamer is not registered and no location information is available or the provide roaming number procedure has failed because of the IMSI detached flag being set).

- *No roaming number available.* A roaming number cannot be allocated because all the available MSRNs in a VMSC are already in use.

- *Facility not supported*

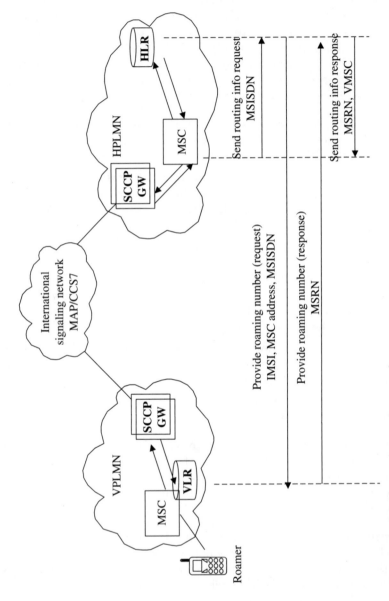

Figure 6-14 Provide roaming number procedure.

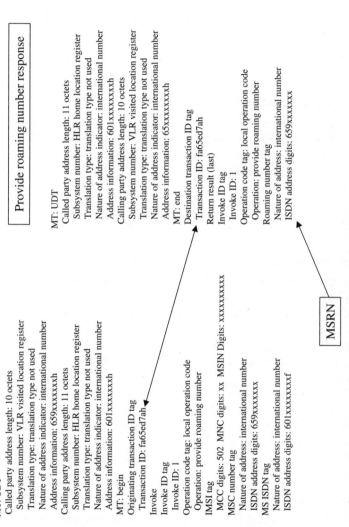

Provide roaming number request

MT: UDT
Called party address length: 10 octets
Subsystem number: VLR visited location register
Translation type: translation type not used
Nature of address indicator: international number
Address information: 659xxxxxxh
Calling party address length: 11 octets
Subsystem number: HLR home location register
Translation type: translation type not used
Nature of address indicator: international number
Address information: 601xxxxxxh
MT: begin
Originating transaction ID tag
Transaction ID: fa65ed7ah
Invoke
Invoke ID tag
Invoke ID: 1
Operation code tag: local operation code
Operation: provide roaming number
IMSI tag
MCC digits: 502 MNC digits: xx MSIN Digits: xxxxxxxxx
MSC number tag
Nature of address: international number
ISDN address digits: 659xxxxxx
MS ISDN tag
Nature of address: international number
ISDN address digits: 601xxxxxxxf

Provide roaming number response

MT: UDT
Called party address length: 11 octets
Subsystem number: HLR home location register
Translation type: translation type not used
Nature of address indicator: international number
Address information: 601xxxxxxh
Calling party address length: 10 octets
Subsystem number: VLR visited location register
Translation type: translation type not used
Nature of address indicator: international number
Address information: 65xxxxxxxh
MT: end
Destination transaction ID tag
Transaction ID: fa65ed7ah
Return result (last)
Invoke ID tag
Invoke ID: 1
Operation code tag: local operation code
Operation: provide roaming number
Roaming number tag
Nature of address: international number
ISDN address digits: 659xxxxx

MSRN

Figure 6-15 Example message decode for provide roaming number.

- *Unexpected data value.* The data type is formally correct but its value or presence is unexpected in the current context.

- *System failure.* The task cannot be performed because of a problem in other entity. The type of entity or network resource may be indicated by the network resource parameter.

- *Data missing.* An optional parameter required by the context is missing.

Figure 6-16 shows the provide roaming response with an error.

6.3.4   Cancel location

When a roamer moves from one VLR area to another area within the PLMN where it was initially roaming or switches to another PLMN, the HPLMN HLR uses the cancel location procedure to inform the old VLR. On receiving this message, the old VLR deletes the roamer's data from its database.

The MAP cancel location request carries the roamer's IMSI (for which this procedure was invoked) as a parameter.

6.3.5   Purge MS

The VPLMN VLR deletes the roamer's record from its database, if a roamer is inactive for a extended period of time, and sends a MAP purge MS request to the HPLMN HLR. On receiving purge MS message from the VLR, the HLR marks (sets the MS purged flag) the MS so that any request for routing information for a mobile-terminated call or mobile-terminated short message will be treated as if the MS were not reachable.

The network provider sets the period after which this procedure should be invoked. This procedure is also invoked in case roamer information needs to be deleted by an operator using a man-machine command.

The MAP purge MS request message carries the roamer's IMSI (for which the procedure was invoked) as a parameter.

6.3.6   Restore data

A VPLMN VLR invokes the restore data procedure in the following scenario;

- On receiving a MAP provide roaming number request from the HPLMN HLR for an IMSI that is not known in the VLR.

- On receiving a MAP provide roaming number request from the HPLMN HLR for a known IMSI for which the confirmed by HLR flag

Provide roaming number request

MT: UDT
Called party address length: 11 octets
Subsystem number: VLR visited location register
Translation type: translation type not used
Nature of address indicator: international number
Address information: yyyyyyyyyh
Calling party address length: 11 octets
Subsystem number: HLR home location register
Translation type: translation type not used
Nature of address indicator: international number
Address information: xxxxxxxxxh
MT: begin
Originating transaction ID tag
Transaction ID: 7f00f5h
Invoke
Invoke ID tag
Invoke ID: 1
Operation code tag: local operation code
Operation: provide roaming number
IMSI tag
    MCC digits: xxx MNC digits: xx MSIN digits: xxxxxxxxx
MSC number tag
Nature of address: international number
ISDN address digits: xxxxxxxxxxf
Nature of address: international number
ISDN address digits: xxxxxxxxx

Provide roaming number response

MT: UDT
Called party address length: 11 octets
Subsystem number: HLR home location register
Translation type: translation type not used
Nature of address indicator: international number
Address information: xxxxxxxxxh
Calling party address length: 11 octets
Subsystem number: VLR visited location register
Translation type: translation type not used
Nature of address indicator: international number
Address information: yyyyyyyyyh
MT: end
Destination transaction ID tag
Transaction ID: 7f00f5h
Return error
Invoke ID tag
Invoke ID: 1
Error code tag: local error code
Error code: absent subscriber

Failed procedure

Figure 6-16 Example message decode for provide roaming number (response).

159

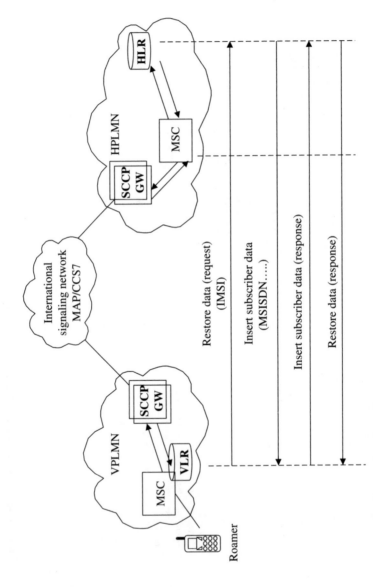

Figure 6-17 Restore data procedure.

is set to not confirmed. This flag is set because the data is assumed to be not reliable after the HLR restart, as a result of severe error.

As shown in Figure 6-17, on receiving a Restore Data message, the HPLMN HLR invokes the insert subscriber data procedure to synchronise the VLR data for a certain roamer/IMSI.

6.4   Roaming Call Scenarios

This section describes voice call scenarios for a roamer in a visited network.

6.4.1   Roamer-originated call

Once the initial authentication and the update location procedures are complete, the visited network allows roamers to use all the services, subject to any restrictions set by the HPLMN. To initiate an outgoing call while visiting a foreign network, a roamer needs to key in the complete international number, including international prefix, i.e.,+<CC><NC> <MSISDN>. If the roamer is allowed to make an outgoing call, the call is processed and controlled by the visited network, which uses its own resources. The HPLMN plays no further role in call processing.

6.4.2   Roamer-terminated call

When an MS roams in a VPLMN, the most likely MT call scenarios are as follows:

1. Caller is an HPLMN subscriber

2. Caller is from the same home country but a different PLMN

3. Caller is a VPLMN subscriber

4. Caller is a subscriber from a country other than that of the HPLMN/VPLMN

Figure 6-18 shows a roamer-terminated call scenario, where the caller is from a country other than that of the HPLMN or the VPLMN. In principle, the call sequence and procedures remain the same. The only difference lies in points of interconnect and the entity to interrogate the HPLMN HLR.
The steps are as follows:

1. A subscriber (country 1) dials the MSISDN of the called MS, which is a subscriber in a PLMN of country 2. The dialing sequence contains the international prefix of country 1 and the full MSISDN of the called MS, including country and network code.

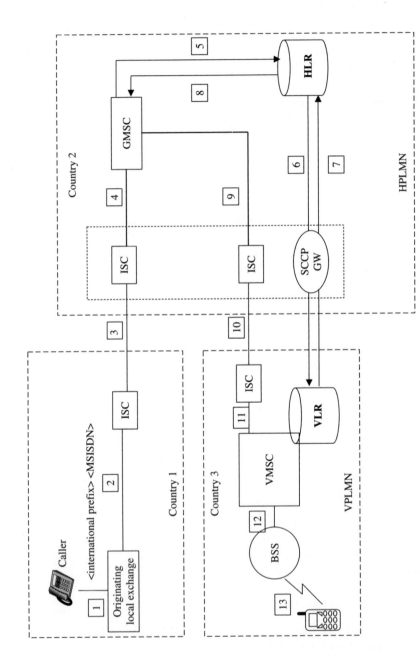

Figure 6-18 Roamer-terminated call.

2. The local exchange analyzes the dialed digits and, on recognizing the international prefix, routes the call to an international switching center (ISC).

3. The ISC analyzes dialed digits further and, on recognising the country code (CC) for country 2, routes the call to the ISC in country 2, using the international routing mechanism.

4. The ISC directs the call to the GMSC in the HPLMN where the called party is subscribing. This routing is done on the basis of network code analysis.

5. The dialed digits received by the GMSC contain no indication of the location of the called MS. In order to route the call to the MS, the GMSC must be aware of the location of the MS and the routing address to be used. The GMSC analyzes the received digits to identify the right HLR to which the query for routing information needs to be sent. The GMSC sends a request to the HLR for the routing information, using the MAP procedure.

6. The HLR checks its data, using MSISDN as a key to identify the VLR where the MS is currently roaming. It sends a message to the VLR to get the roaming number assigned to the MS. In this example, as the MS is roaming in a different PLMN in a foreign country, the request is routed to the VPLMN using an international SCCP connection.

7. On receiving the request, the VLR assigns a temporary number, i.e., MSRN, to the roaming MS for the routing purpose and sends the response back to the requesting entity, i.e., the HPLMN HLR.

8. The HLR passes the received MSRN to the GMSC.

9. The GMSC analyzes the MSRN and routes the call to the ISC, using ISUP call procedures.

10. The ISC analyzes the country code in MSRN and directs the call to the ISC in country 3.

11. The ISC in country 3 analyzes the MSRN further to identify the PLMN and route the call to the VMSC (or GMSC, which in turn routes the call to the VMSC).

12. The VMSC then routes the call by the usual call control procedures.

As can be seen, this call involves two international call legs. This is not an efficient call routing. One call leg back to the home country can be saved if the calling network (country 1, in this case) is allowed to interrogate the HPLMN HLR.

6.5   Short Message Service (SMS)

Short message service is made up of two basic services:

- SM MT   (short message, mobile terminated)
- SM MO   (short message, mobile originated)

Figure 6-19 shows the concept of SM MO and SM MT for a roamer.

SM MT denotes the capability of the GSM system to transfer a short message submitted from the short message service center (SMSC or just SC for short) to a subscriber's mobile station. It also provides information about the delivery of the short message, either by a successful delivery report or a failure report with a specific mechanism for later delivery.

SM MO denotes the capability of the GSM system to transfer a short message submitted by a subscriber mobile station to another subscriber via an SC. It also provides information about the delivery of the short message, either by a successful delivery report or a failure report. The message must include the MSISDN number of the destination subscriber to which the SC will attempt to relay the short message.

In early versions of MAP (i.e., MAP v2 and lower) the forward short message procedure is used for both mobile-originated and mobile-terminated short messages. The direction (i.e., MO or MT) is determined by examining the SM-PR-DA parameter, which in the case of MT SMs contains the IMSI of the receiver.

In later versions of MAP, SM MTs and SM MOs are handled separately by means of the MT forward short message and MO forward short message procedures.

6.5.1   Roamer-originated SM

When a roamer in a foreign network submits short message (SM), the serving MSC invokes the MAP MO forward short message procedure toward the SMS interworking MSC (IWMSC) in the HPLMN, which forwards it to an SMSC. Figure 7-20 shows the MAP v2+ procedure for an SM originated by a roamer.

The following parameters are sent from the serving MSC to the IWMSC

- *SM RP DA.*   This parameter represents the destination address. For a roamer-originated SM, it contains the service center address received from the mobile station.

- *SM RP OA.*   This parameter represents the originating address. For a roamer-originated SM, it contains the MSISDN of the sending party.

- *SM RP UI.*   This parameter represents the user data field carried by the short message service, i.e., transfer protocol data unit.

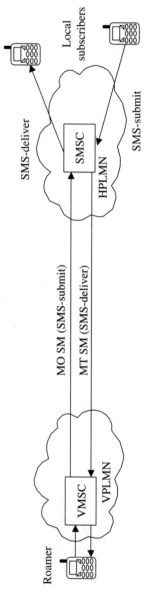

Figure 6-19 SM MO and SM MT to a roamer.

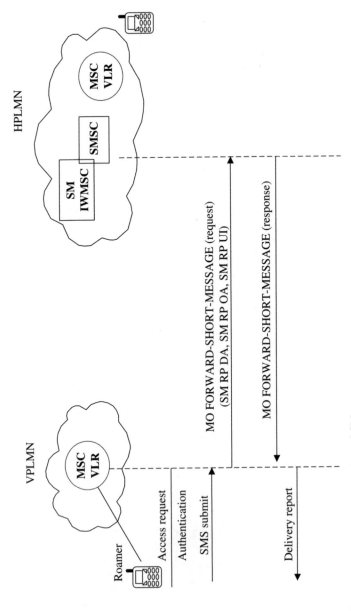

Figure 6-20 Roamer-originated SM.

The following error messages (according to the nature of fault) are returned by the IWMSC when the service fails -

- *Facility not supported.*   The requested facility is not supported by the PLMN.

- *System failure.*   The task cannot be performed because of a problem in another entity. The type of entity or network resource may be indicated by a network resource parameter.

- *SM delivery failure.*   The reason of the SM delivery failure can be one of the following in the mobile-originated SM:
  - Unknown service center address
  - Service centre congestion
  - Invalid short message entity address
  - Subscriber not service center subscriber
  - Protocol error
  - Unexpected data value

6.5.2   Roamer-terminated SM

When a HPLMN SMSC receives a short message to be relayed, the SMSC sends the short message to the SMS-GMSC, which interrogates the HLR to retrieve the routing information (i.e., serving MSC address) needed to forward the short message. The SMS GMSC then sends the short message to the relevant MSC, transiting other networks as required. The serving MSC then delivers the short message to the roamer.

Figure 6-21 shows the MAP v2+ procedure for a roamer-terminated SM.

The MAP send routing info for SM procedure is used between the gateway MSC and the HLR to retrieve the routing information for routing an SM to the MSC currently serving the roamer. The following parameters are sent to the HLR:

MSISDN.   The destination MS address.

SM-RP-PRI.   This parameter indicates the priority of the short message. It is used to determine whether delivery of the short message will be attempted when a service center address is already contained in the message waiting data file.

Service center address.   The address of the sender SMSC.

The MAP MT forward short message procedure is used between the gateway MSC and the serving MSC to forward mobile-terminated short messages.

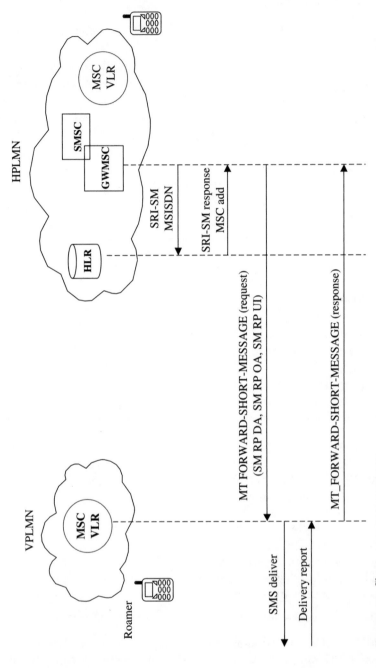

Figure 6-21 Short message to a roamer.

The following parameters are sent from the serving MSC to the IWMSC:

- *SM RP DA.* This parameter represents the destination address. For a roamer-terminated SM, it contains the IMSI of the MS to which the SM is being sent.
- *SM RP OA.* This parameter represents the originating address. For a roamer-terminated, SM it contains the originating SMSC address.
- *SM RP UI.* This parameter represents the user data field carried by short message service, i.e., the transfer protocol data unit.

The errors related to this procedure are:

- *Unknown subscriber.* The PLMN has rejected the short message because an IMSI or a directory number for the mobile subscriber is not listed in the HLR.
- *Absent subscriber.* The PLMN has rejected the short message because there was no paging response from the MS.
- *Subscriber busy for MT short message.* The PLMN has rejected the short message because congestion was encountered at the visited MSC. Possible reasons for this include any of the following events in progress-
  - Short message delivery from another SC
  - Location update
  - Paging
  - Emergency call
  - Call setup
- *Facility not supported.* The VPLMN has rejected the short message because there is no provision for the SMS in the VPLMN
- *Illegal subscriber.* The PLMN has rejected the short message because the MS failed authentication.
- *Illegal equipment.* The PLMN has rejected the short message because the IMEI of the MS was blacklisted in the EIR.
- *SM delivery failure.* The reason of SM delivery failure can be one of the following:
  - Protocol error
  - MS does not support SM MT
- *Unexpected data value.* The data type is formally correct but its value or presence is unexpected in the current context.
- *System failure.* The task cannot be performed because of a problem in another entity. The type of entity or network resource may be indicated by a network resource parameter.

- *Data missing.* An optional parameter required by the context is missing.
- *Memory capacity exceeded.* The MS rejects the short message because it has no memory capacity available to store the message.

6.5.3   MAP v2 procedures

MAP v2 and lower versions use the MAP forward short message procedure for both roamer-originated and -terminated SMs from the interworking MSC (IWMSC) to the serving MSC. The following parameters are sent from the IWMSC to the service MSC:

- *SM RP DA.* When a MT SMS is being forwarded, this parameter contains IMSI or LMSI. When a MO SMS is being forwarded, this parameter contains the service center address as received from the MS.
- *SM RP OA.* When an MT SMS is being forwarded, this parameter contains the SMSC address. When an MO SMS is being forwarded, this parameter contains the MSISDN of the sending party.
- *SM RP UI.* This parameter indicates that the short message transfer protocol data unit contains actual data (i.e., a short message) from the MS.
- *More messages to send.* This parameter indicates that more messages in the SMSC are to be sent to the destination.

Bibliography

ITU-T Recommendation E.164, The international public telecommunication numbering plan.
ITU-T Recommendation E.212, The international identification plan for mobile terminals and mobile users.
ITU-T Recommendation E.213, Telephone and ISDN numbering plan for land Mobile Stations in public land mobile networks (PLMN).
ITU-T Recommendation Q.713, Specifications of Signalling System No.7; SCCP formats and codes.
ITU-T Recommendation Q.716, Specifications of Signalling System No.7; Signalling connection control part (SCCP) performances.
3GPP TS 29.002, Mobile Application Part (MAP) specification.
3GPP TS 22.090, Unstructured Supplementary Service Data (USSD)—Stage 1.
GPP TS 23.012, Location registration procedures.
3GPP TS 23.040, Technical realization of the Short Message Service (SMS) Point to Point (PP).
3GPP TS 23.003, Numbering, addressing and identification.
GSM 03.04, Signalling requirements relating to routing of calls to Mobile Subscriber.

Chapter

7

Roaming in GPRS and 3G Networks

At the time of writing, many GPRS and 3G networks have already been successfully deployed. However, deployment of international roaming is still work in progress. Users take GSM roaming for granted and expect GPRS/3G services to roam just the same. The data roaming capabilities of GPRS/3G give users fast and easy access to the information while they are roaming in networks other than that of their home service provider. Email, home location stock market reports, office intranet, personal banking and a host of WAP services are just a few examples. Data roaming is fundamental to making future global mobile Internet services a reality.

3G technology requires major changes in the access network to support applications, which require high data rates. To start with, 3G implementation with Rel 99 uses the same basic architecture as GPRS in the core network. Most of the existing core network elements, such as the SGSN and the GGSN, will be part of the 3G networks after upgrading. The general principles on which roaming in 3G is implemented are no different from those already established for GSM (voice) and GPRS (data). The illustrations and descriptions in the following sections are based on data roaming in GPRS networks, but can be generally applied to 3G networks.

7.1 Inter-PLMN Connectivity

Roaming implementation in GPRS/3G is not a simple extension to GSM voice roaming. As shown in Figure 7-1, a new inter-PLMN IP backbone is required to interconnect PLMNs to enable packet data transfer.

As is the case for the GSM, CCS7/SCCP connectivity between PLMNs must exist for MAP signaling exchange. For GPRS/3G roaming, MAP Version 3 (also referred as MAP 2+) is mandatory. Older MAP versions

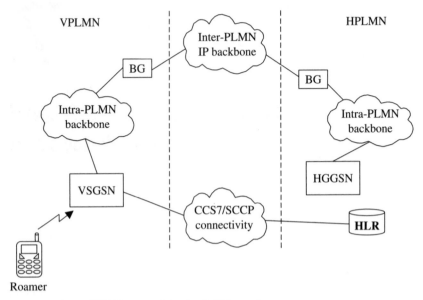

Figure 7-1   Inter-PLMN connectivity for GPRS roaming.

do not support GPRS-specific procedures. The operators that already have GSM roaming in place may use the same connectivity for GPRS/3G.

So the first step is to establish IP connectivity between partner networks for IP data routing, i.e., the IP network interconnecting GSNs and the inter-PLMN backbone network of different PLMNs. This connectivity did not exist before the advent of GPRS.

Figure 7-2 shows inter-PLMN connectivity requirements for 3G roaming. For 3G, inter-PLMN connections include:

- CCS7 connectivity to transfer MAP and ISUP signaling
- Inter-PLMN IP backbone for packet data routing
- International voice trunks between partner networks

The infrastructure build for GSM and GPRS roaming can be leveraged to deploy 3G roaming.

7.1.1   Inter-PLMN backbone network

GPRS specifications support a connection between two PLMNs by using the Gp interface. The Gp interface is defined in order to make services of the HPLMN also available for the roamers in the VPLMN. The inter-PLMN backbone network is required to create GTP tunneled PDP contexts between GSNs in different PLMNs. Moreover, the inter-PLMN backbone

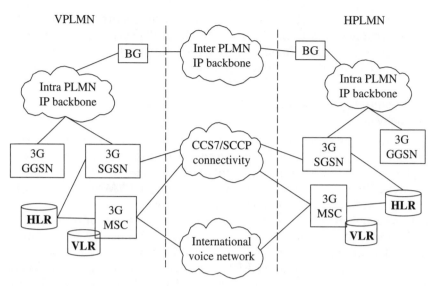

Figure 7-2 Inter-PLMN connectivity for 3G roaming.

network is also needed to enable interworking of other nodes such as MMSCs in different PLMNs. The Gi interface is defined between a PLMN and an external data network. The external data network may be public, such as the Internet, or private corporate networks. Figure 7-3 illustrates the connection between PLMNs and external data networks.

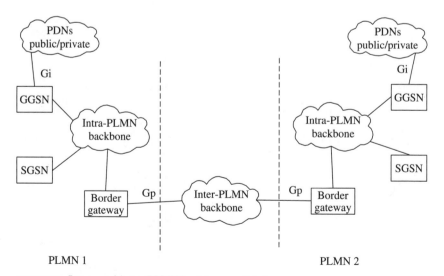

Figure 7-3 Intra- and inter-PLMN backbone network.

Border gateways (BGs) are used to provide secure links between PLMNs connected via the Gp or Gi interface. BGs typically consist of a router and a firewall. Roaming traffic (i.e., data and signaling between GSNs) is carried by using the GTP protocol.

7.1.2 Inter-PLMN data connectivity alternatives

Currently, there are more than 400 GSM operators. Most are migrating to support GPRS- and 3G-based services. It is a complex and costly task to provide data interconnectivity between one PLMN and all other PLMNs. There are several ways to interconnect PLMNs, as shown in Figure 7-4, for the enabling data packet transfer.

It is feasible to connect GPRS operators by direct links for the purposes of limited trials or service testing. However, this is not a practical option for the long term. Direct connectivity between GPRS operators is achieved by using direct leased lines or VPN-based leased lines. Another option is to use tunneling via the public Internet. Direct leased lines and VPN provide guaranteed QoS and are more secure, but are expensive options. Using the public IP network is the most economical option, but it compromises the security of the GPRS network and prohibits offering guaranteed high QoS levels, as would likely be required for corporate user service level agreement (SLA) commitments. However, operators use this option initially for trials and jump starting services. It is also an attractive option for the smaller and newer operators.

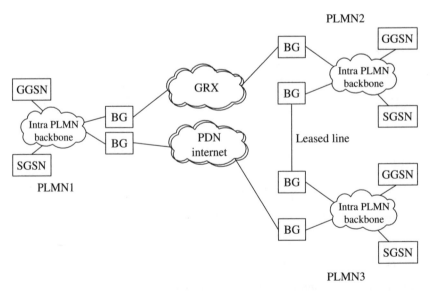

Figure 7-4 Inter-PLMN backbone infrastructure options.

The two options discussed here may be practical for limited deployment. However, these are not suitable options to implement roaming on a global basis. The GSM Association proposal on GPRS Roaming eXchange (GRX) is a long-term solution for inter-PLMN IP connectivity. The next section covers more details on GRX concepts and implementation.

Figure 7-4 illustrates three options. PLMN 1 is shown connected to PLMN 3 by a public packet data network (the Internet). PLMN 2 and PLMN 3 are connected directly by leased line. PLMN 1 and PLMN 2 are connected through a GRX operator.

Border gateways and firewalls ensure the security of GPRS PLMNs from unsolicited traffic and signaling from other PLMNs and from fraudulent sources. Internetwork secure tunneling and encryption is achieved by using BGs.

7.1.3   GPRS roaming eXchange

The GSM Association proposed a global GPRS roaming network for inter-PLMN connectivity as a long-term solution. This proposal eliminates the need for dedicated connections between every roaming partner and enables GPRS service providers to offer roaming service by using just one connection to the global roaming network. The global GPRS roaming network is made up of GPRS Roaming eXchange (GRX) nodes connected to each other directly or through other GRXs capable of transiting traffic between GRX nodes. Figure 7-5 illustrates the general architecture of the GRX as

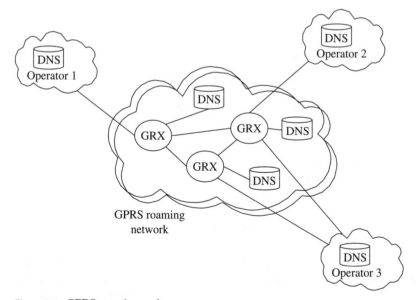

Figure 7-5   GPRS roaming exchange.

proposed by the GSM Association. A PLMN can be connected to a GRX with a Layer 1 connection, using leased lines or fiber links; a Layer 2 logical connection, using ATM, LAN, or Frame Relay; or a Layer 3 IP VPN connection, using the public Internet.

The GRX infrastructure is built and operated by separate service providers. Many established cellular service providers and international data carriers have already launched commercial GRX services that eliminate the need for establishing direct connections between the PLMNs. The GSM Association has outlined general requirements for GRX service providers.

The following services must be offered by GRX service providers:

- Physical connectivity to PLMNs

- IP addressing for the inter-PLMN backbone

- Routing of data packets and DNS queries between a PLMN and its roaming partners (supporting GTP tunneling on both UDP and TCP, as required by GTP specifications)

- DNS root services

- Dynamic exchange of routing information between connected GPRS networks using BGP-4 routing protocol capabilities

- Data security against spoofing and intrusion from the public networks

- Service level guarantees

There are two possible connection scenarios for GPRS PLMNs over GRX:

- Direct connection, where two GPRS operators are connected by a single GRX

- GPRS peering, where two GPRS operators are connected over two or more GRX operators (used when the preferred GRX provider does not have a global presence, for example)

As shown in Figure 7-6, the PLMN 1 and the PLMN 2 are directly connected using GRX operator A. PLMN 1 and PLMN 3 are connected by GPRS peering. PLMNs need to have point-to-point roaming agreements for use of either direct or peering GRX connections.

7.1.4 Border gateway protocol version 4

The border gateway protocol (BGP) is an inter-autonomous system (AS) routing protocol used for the global Internet. An AS is a set of routers under a single technical administration that uses an interior gateway protocol and common metrics to route packets within the AS. BGP-4 is

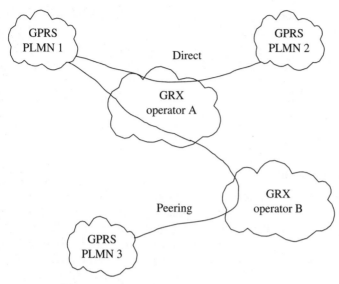

Figure 7-6 Direct connection and GRX peering.

an exterior gateway protocol used to route packets to other ASs. The primary function of BGP peers is to exchange complete copies of AS routing tables. As routing tables of AS can be very large, BGP exchanges an entire routing table on the initial connection. Later connections exchange only incremental updates. This makes BGP sessions more efficient. TCP is the transport protocol used by the BGP.

In Global GPRS networks, each PLMN is treated as an AS. GRX Operators need to support the BGP-4 routing protocol to advertise all known routes to GPRS operators, so that a completely meshed roaming network is provided.

7.2   Access Point Name

GPRS networks create a logical connection between the MS and an external PDN by using an access point name (APN). The APN indicates which GGSN in the GPRS backbone network is to be used. At the GGSN, it may further indicate the external data network or services to which the MS should be connected.

A list of allowed APNs for each subscriber is stored in the HLR as a part of subscription data. The SGSN compares the APN received by a roamer in an **activated PDP context** message with subscriber data in the HLR to check whether the requested service is authorized. DNS functionality is used to translate the APN to the GGSN IP address.

An APN is constructed in the same way as domain names are on the Internet. The syntax is as follows:

my.isp.com.mnc<MNC>.mcc<MCC>.gprs

An APN consists of two parts:

- APN network identifier
- APN operator identifier

7.2.1 APN network identifier

The APN network identifier indicates the external PDN network that the GGSN is connected to and the service the subscriber wishes to access. A network identifier is a mandatory parameter. Each GGSN in a different PLMN is given a unique identifier to avoid address conflicts.

Typically, an APN network identifier follows the Internet URL coding format, consisting of three or more labels, starting with the reserved service label. This name is allocated by the PLMN to an organization that has officially reserved the name in the Internet domain.

However, an APN may be just one label corresponding to a DNS name of a GGSN. This is locally interpreted, by the GGSN as a request for a specific service. Examples of APN network identifier are:

my.isp.com

ptt

MMS

The network identifier is provided by a GPRS user or predefined in a GPRS terminal.

7.2.2 APN operator identifier

An APN operator identifier is an optional part of the APN. The APN operator identifier is placed after the APN network identifier, if present. The APN operator identifier consists of three labels in the following format:

mnc<MNC>.mcc<MCC>.gprs

where MNC and MCC are derived from the subscriber IMSIs. The last label is gprs by default.

The APN operator identifier is not stored in the HLR as part of the subscription data. It can be input by the users or inserted by an SGSN.

The default APN operator identifier is used in inter-PLMN roaming situations in attempting to translate an APN consisting only of a network identifier into the IP address of the GGSN in the HPLMN.

7.2.3   Wild card APN

Wild card APN, i.e., "*," is set in the HLR subscription data to indicate that the HPLMN operator allows the roamer to access any network of a given PDP type. On receiving an **activate PDP context** message from the MS the subscription data of which is set to wild card APN, the serving SGSN either chooses to use the default APN network identifier or an APN network identifier received in the request message for addressing the GGSN.

7.3   APN Resolution

7.3.1   GPRS and DNS

Domain name system (DNS) functionality is used for mapping logical names to IP addresses in the GPRS intra- and inter-backbone network. Each PLMN has its own local DNS functionality. Typically, two mirrored DNS servers, i.e., primary and secondary DNSs, are available to ensure uninterrupted service. In addition, the inter-PLMN backbone also has DNS functionality, i.e., root DNS. SGSNs communicate with their own local network DNS for mapping. The DNS of different PLMNs talk to each other either directly or through a DNS root maintained by GRX operators or a master DNS maintained by the GSM Association.

Figure 7-7 shows the topology and communication flow of DNS systems for inter-PLMN mapping. The master root DNS server holds the records of the responsible domain of each operator's DNS server. Each GRX operator

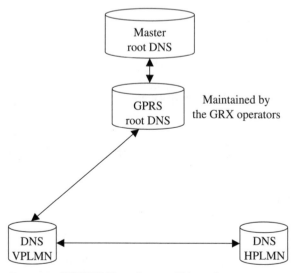

Figure 7-7   GPRS DNS topology and hierarchy.

periodically synchronizes with the master root DNS to its own root DNS server. DNS to DNS communication uses IP to transfer information.

Note that the GPRS DNS system is a private network. It has no interaction with the Internet's DNS system.

7.3.2 APN resolution using DNS in the HPLMN

In this case, a roamer is attached to a VPLMN SGSN (VSGSN) and activates a PDP context, using a GGSN in the home network. The VSGSN needs to know the HGGSN address to initiate the create PDP context procedure. The VSGSN uses an APN provided by the user to resolve IP addresses with the help of DNSs. Figure 7-8 illustrates APN resolution procedure in detail.

The procedure consists of these steps:

1. A roamer currently attached to the GPRS network activates a GPRS service. The MS sends an **activate PDP context** message to the serving SGSN. This message may or may not contain an APN name corresponding to the services the user wishes to access. Let us assume that the APN is my.isp.com.

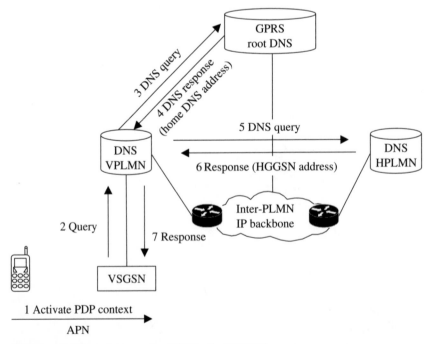

Figure 7-8 APN resolution using DNS in the HPLMN.

2. The VSGSN inserts its own operator identifier (e.g., mnc011.mcc111.gprs) to make a complete logical name, i.e., my.isp.com.mnc011.mcc111.gprs, and queries its own local DNS. If the APN my.isp.com is not known to the local DNS, it responds back indicating failure to resolve the APN.

3. The VSGSN now inserts the roamer's home operator identifier' (e.g., mnc022.mcc222.gprs) and sends a query to the root DNS (via its local DNS) in the inter-PLMN backbone.

4. The GPRS root DNS replies by sending the HPLMN DNS address to the VPLMN DNS.

5. The VPLMN DNS asks the HPLMN DNS for the GGSN address.

6. The HPLMN DNS resolves the APN and responds back to the VPLMN DNS.

7. The VPLMN DNS replies to the VSGSN with the HGGSN address.

8. The VSGSN initiates the **create PDP context** procedure with the HGGSN.

7.3.3 APN resolution using DNS in the VPLMN

In this case, a roamer is attached to a VSGSN and activates a context using a VGGSN. Figure 7-9 shows the steps involved in resolving the APN:

1. A roamer currently attached to the GPRS network activates a GPRS service. The MS sends an **activate PDP context** message to the serving SGSN. This message may or may not contain APN name corresponding to the services the user wishes to access. Let us assume that the APN is my.isp.com.

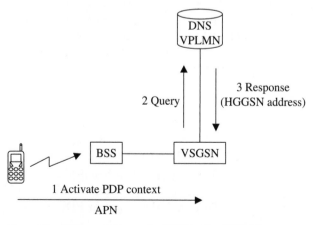

Figure 7-9 APN resolution using DNS in the VPLMN.

2. The VSGSN inserts its own operator identifier (e.g., mnc011.mcc111. gprs) to make a complete logical name, i.e., my.isp.com.mnc011mcc111. gprs, and queries its own local DNS.

3. The APN my.isp.com is known to the local DNS. It responds back with the VGGSN address, which is required to serve the roamer request.

4. The VSGSN then activates the **create PDP context** procedure with the GGSN.

7.4 Roaming Scenarios

7.4.1 GPRS attach in a visited network

The roamer registers itself in a visited network, using the GPRS Attach procedure. The visited SGSN (VSGSN) communicates with the home network HLR to perform authentication, update the location, and get the roamer's subscription data. When a roamer attaches for the first time in the visited network, the only information available to the SGSN for it to route a signaling message to the roamer's home network is the IMSI. The SGSN drives the mobile global title (MGT) from the IMSI as described in Chapter 6.

Figure 7-10 shows the GPRS Attach procedure. The steps are as follows:

1. The MS sends a GPRS **attach request** with its identity, i.e., IMSI, and other necessary information such as the old routing area indicator (RAI) to the VSGSN.

2. The VSGSN initiates the authentication procedure with the HPLMN HLR. The MAP protocol is used for communication between the VSGSN and the HLR. The VSGSN may also request MS to identify itself by invoking identification request procedure.

3. On successful authentication, the VSGSN initiates the update location procedure. The update GPRS location procedure is used by the SGSN to update the location information stored in the HLR. It uses the following mandatory parameters.

   - IMSI
   - SGSN number
   - SGSN address

   The SGSN number refers to the ISDN number of an SGSN. The SGSN address refers to the IP address of an SGSN.

4. The HLR sends subscription information to the VSGSN using the **insert subscriber data** procedure. The HLR updates the SGSN with GPRS subscription data. This includes a list of PDP contexts that the subscriber has access to.

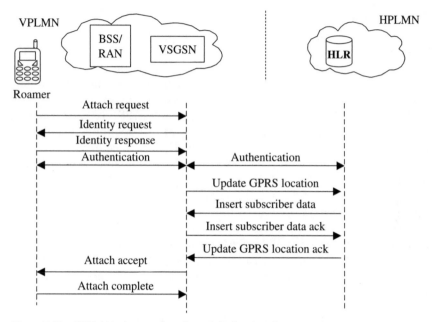

Figure 7-10   GPRS Attach procedure in a visited network.

5. The VSGSN indicates a successful attach by sending an **attach accept** message.

6. The MS acknowledges with an **attach complete** message.

   In case of GPRS attach failure, one of the following errors is returned in an attach reject message:

- Illegal MS
- GPRS service not allowed
- GPRS and non-GPRS services not allowed
- PLMN not allowed
- Location area not allowed
- Roaming not allowed in this location area
- GPRS services not allowed in the PLMN
- No suitable cells in location area

   After a successful GPRS attach, the MS is in the ready state and mobility management (MM) contexts are established in the MS and the VSGSN. The MS can send and receive SMS at this stage.

7.4.2   PLMN and ISP roaming

The case where the roamers in a visited network use a gateway in their home network to access services is often referred as PLMN roaming/Gp roaming. The advantage of this method is that roamers are fully transparent to their location. For example, they can access same ISP and portals transparently, as they are in their own network. It also allows collection of charging data in the HPLMN. An international GPRS IP backbone link is required between PLMNs to implement PLMN roaming. The data is exchanged between two PLMNs across the Gp interface.

The case where roamers visited network access data services using VGGSN is often referred as ISP roaming. ISP roaming provides optimal and efficient routing of data. No international GPRS IP backbone link is required in this scenario, as no traffic is routed back to the home network. The HPLMN has only limited control over its own subscribers. The QoS experienced by roamers will largely depend on the visited network.

To begin data transfer, the roamer needs to activate the PDP context. The VSGSN will then determine the applicable roaming scenario according to the following:

1. Selection based on VPLMN address allowed (VPAA flag): The home operator sets a flag in the HLR subscriber data to indicate if a subscriber is allowed to use the visited network's GGSN (VGGSN). The network operator can force its subscribers to use the home GGSN by disabling VPAA in the HLR on a per APN basis. In this case the roamers have no choice but to use HGGSN for access.

2. Selection based on User data: The roamers in the visited network can select their HGGSN by requesting the APN operator identifier of their home operator. The APN contains the user's and network's desired routing access preference and is used to create a logical connection between the user's terminal and external packet data networks. Each PLMN has a primary and secondary DNS server for APN resolution. If the DNS in the visited network is unable to resolve the APN, then it will query the ".gprs" root DNS.

3. In the case where the VPAA flag does not restrict a user from using the visitor network but the VPLMN is not connected to the requested ISP, the VPLMN still offers services through the HPLMN.

7.4.3   PDP context activation using HGGSN

The first step for a roamer to use GPRS/3G data services is to attach to the visited network. The GPRS attach procedure is described in Section 7.4.1. On successful attach, the roamer can send and receive SMSs. To access data services like Internet, WAP, and MMS, the roamer needs to activate the PDP context. As described in the previous section, the PDP context activation is achieved through a HGGSN or a VGGSN. Figure 7-11 illustrates the steps to perform PDP context activation by using an HGGSN.

The steps to establish the PDP context are as follows:

1. The roamer wishes to use the HPLMN-specific APN. The MS sends an **activate PDP context** message to the VSGSN in the VPLMN.

2. The VSGSN initiates APN resolution procedures as described in Section 7.3.2. The APN given by the user is used as the key.

3. As a result of successful APN resolution, the VSGSN gets the IP address for the GGSN in the roamer's home PLMN.

4. The VSGSN sends a **create PDP context** request to the HGGSN.

5. For a successful create PDP context, a response is sent from the HGGSN to the VSGSN with a cause value of request accepted. The SGSN then activates the PDP context and may start to forward packet data units (PDUs) to/from the MS from/to the external data network.

The data packet transfer and involvement of various network elements is shown in Figure 7-12. As shown in the figure, the packet data flows between the MS and the external data network, using GGSN in the home network.

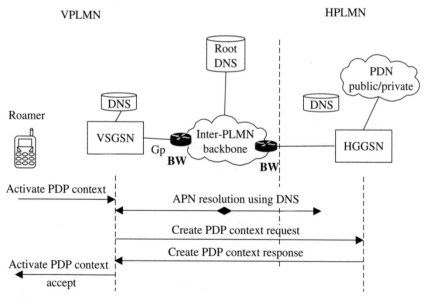

Figure 7-11 PDP context activation using HGGSN.

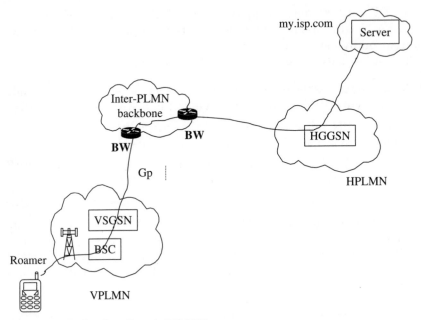

Figure 7-12 Packet data flow via HGGSN

One of the following cause values is returned in the create PDP context response when the context fails to be established:

- No resources available. This indicates that all available IP addresses with the GGSN are occupied.
- Service not supported. This indicates that the GGSN does not support the requested PDP type.
- User authentication failed. This indicates that the external PDN has rejected the service requested by the user.
- System failure.
- Mandatory information element is missing.
- Mandatory information element is incorrect.
- Optional information is incorrect.
- Invalid message format.
- Version not supported.

7.4.4   PDP context activation using VGGSN

Figure 7-13 illustrates the steps to perform PDP context activation using VGGSN.

1. A roamer wishes to access a network using an APN. The MS sends an **activate PDP context** message to the VSGSN in the VPLMN.

2. The VSGSN queries its own PLMN DNS, i.e., the VPLMN DNS for the GGSN IP address using the APN as the key. The procedure for APN resolution using DNSs is described in Section 7.3.3.

3. The VPLMN DNS resolves the APN and provides the VSGSN with the IP address of a VGGSN.

6. The VSGSN sends a **create PDP context** request to the VGGSN.

7. For a successful **create PDP context**, a response is sent from the VGGSN to the VSGSN with a cause value of request accepted. The SGSN then activates the PDP context and may start to forward packet data units (PDUs) to/from the MS from/to the external data network.

The data packet transfer and involvement of various network elements is shown in Figure 7-14. As shown in the figure, the packet data flows between the MS and the external data network, using a GGSN in a visited network. The home network has no further role to play.

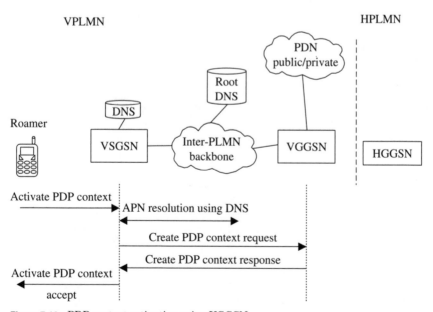

Figure 7-13 PDP context activation using VGGSN.

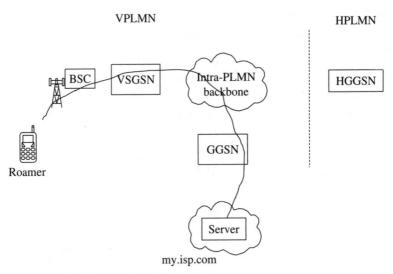

Figure 7-14   Packet data flow via VGGSN.

Bliography

GSM 03.60, Digital cellular telecommunications system (Phase 2+); General Packet Radio
Service (GPRS); Service Description; Stage 2.
GSM 03.03, Digital cellular telecommunications system (Phase 2+); Numbering, address-
ing and identification.
3GPP TS 09.60, GPRS Tunneling Protocol (GTP) across the Gn and Gp interface.
GSM Association PRD IR.33, GPRS Roaming Guidelines.
GSM Association PRD IR.34, Inter PLMN Backbone Guidelines.
RFC 1777, Border Gateway Protocol 4.

8

Roaming Implementation for Prepaid

The prepaid mobile market is evolving all over the world, with prepaid making up a significant and growing number of subscribers. For many operators, prepaid now represents up to 80 percent or more of their subscriber base.

As no credit checks are required, signing a prepaid service is a simple and hassle-free process for new subscribers. For low-income groups like students, prepaid offers new convenience and flexibility in controlling communication expenses. They pay up-front and top up their credit as they use.

Mobile operators are able to expand their subscriber base significantly by offering prepaid services. The possibilities of fraud and bad debts are minimal because of the up-front payment method. However, the churn rate is very high, and gaining the loyalty of prepaid subscribers poses a great challenge for mobile operators.

Extending roaming capabilities for prepaid subscribers to match their postpaid counterparts makes business sense for mobile operators. For prepaid subscribers, an additional verification needs to be done to ensure that an adequate credit balance is available before allowing them to use the services. Moreover, the call needs to be monitored during the conversation period, in real time, to avoid a negative balance.

There are many ways to implement prepaid roaming. Currently, most of the prepaid roaming implementations utilize the capabilities provided by unstructured supplementary service data (USSD) and customized application mobile enhanced logic (CAMEL).

USSD deploys a callback mechanism, as described in the following section. For each outgoing call initiated by a roamer, there will be two call legs—an international call leg back home and a follow-on call. This is surely not an efficient way to handle the call. The offered services are also not transparent, as the user needs to initiate the service by using a special service code. However, implementing prepaid roaming with USSD is rather simple, fast, cheaper, and supported by almost all existing networks, providing a global footprint.

The CAMEL-based solution is technically more advanced and efficient. The user access services transparently, as normal in his home network. There is no additional call leg required. However, the partner networks must also support CAMEL to enable roaming.

8.1   Prepaid Roaming Using USSD Callback

The most popular technology that enables roaming for prepaid subscribers uses unstructured supplementary services data (USSD) capability, which is already built into GSM standards. USSD provides session-based messaging between mobile terminals and mobile services and can also be used to enable many other value-added services such as WAP, interactive chat, prepaid balance checks, and voucher top-up.

USSD uses the callback principle to enable prepaid roaming. A roamer in a visited network requests a call to be set up by keying a special service or access code and the MSISDN number of the destination party. The visited network passes this service request to the roamer's home network. From this point onward, the HPLMN takes control of call processing. For example, the HPLMN may like to perform precall checks before processing further. It verifies if the subscriber has enough credit balance to pay for the requested service. After verification, the prepaid roaming application in the HPLMN initiates a call to the destination subscriber and then initiates another call to the roamer. Once the call is established between roamer and intended called party, the HPLMN monitors the service usage against the available balance in real time. The HPLMN notifies the roamer if the credit balance runs out and terminates the call.

Figure 8-1 shows a typical sequence of an outgoing call initiated by a roamer in a visited network. In this example, the PLMN 2 uses the service code 111 for prepaid roaming, a roamer (PLMN 2 subscriber) in the visited network (PLMN 1) initiates a call request to a subscriber (e.g., +60133439128), which may belong to an HPLMN, any other PLMN, or a fixed network, by sending the following sequence.

111<Called party MSISDN/ISDN>#Send

1. The roamer keys in *111*+60133439128# and presses the <SEND/OK> button to initiate a call to subscriber +60133439128 in another network.

2. The VPLMN MSC/VLR transfers this USSD string to the HPLMN HLR.

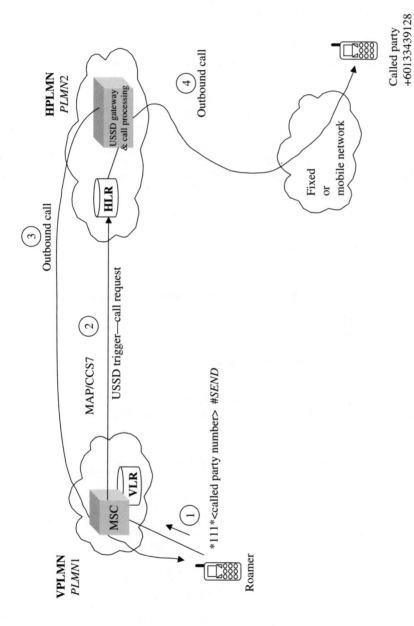

Figure 8-1 USSD Call back—call flow.

3. The HPLMN HLR passes this string to the prepaid roaming platform.

4. After the required precall checks, the prepaid roaming platform initiates two outgoing calls (i.e., one to the roamer and one to the requested called party) and connects.

8.1.1 USSD string coding

The length and content of a USSD string is very flexible. USSD utilizes the characters "∗," "#," and all the digits. If the user keys in a string following the coding scheme given in Table 8-1, the USSD handlers in the MS, the MSC, the VLR, and the HLR treat it as a USSD request.

The character "∗" is used to indicate the beginning of a USSD string. It is also used as a separator between two parameters. The character "#" is used to terminate a USSD string.

The formatting rules to create a USSD string are summarized as follows.

- An asterisk to indicate beginning

- A predefined access/service code

- One or more parameters needed to invoke supplementary service

- A # to terminate request

For example, a USSD service request is coded as follows.

∗<access code>∗<called party number>#

8.1.2 USSD request handling—general concept

As described in Table 8-1, the treatment of USSD at every node is independent of any application in that particular node. A USSD handler

TABLE 8-1   USSD Operations—Interpretation at VPLMN

USSD string	Treatment	Action at the VPLMN	Remarks
∗1XY∗<any number of any characters># where X = 0–4 and Y = 0–9	Reserved for the HPLMN	The VPLMN passes theUSSD message directly to the HPLMN	String may begin with 1, 2, or 3 digits from the set ∗, #
∗1XY∗<any number of any characters># where, X = 5–9 and Y = 0–9	Reserved for the VPLMN	It is up to the VPLMN to decide how to treat it.	
7(Y) where Y = any number 0–9	Reserved for the HPLMN	The VPLMN passes the USSD message directly to the HPLMN	

routes the USSD to the correct application based on service code. Figure 8-2 shows the routing of a USSD message to the destination node.

A roamer in a visited network can initiate a USSD request for a provisioned service at any time by keying in the USSD string containing the HPLMN service code. The MSC examines the service code and forwards this request to the VLR without taking any action.

When a VLR receives the request forwarded by the MSC with an HPLMN service code, it sets up a connection to the HPLMN HLR and forwards the request unchanged. The HLR, on receiving the USSD request, processes and passes it to an appropriate application, i.e., a prepaid roaming application in this case.

In a similar fashion, the network can send a USSD operation toward an MS anytime. This operation could be a request to get certain information or a notification. For example, if the prepaid roaming application in the HLR is to send a USSD request or notification to the roamer, it sets up a transaction to the VLR where the roamer is currently registered. The VLR, on receiving this request or notification, sets up a transaction to the serving MSC. The serving MSC sends it to the MS and then waits for a response. The MS analyzes the data-coding scheme and decides whether the USSD operation is in the MMI mode or the application mode. For a USSD request with the application mode, the MS passes the message to the corresponding application and sends back the response. For a USSD request with the MMI mode, the MS displays the text provided and waits for the user inputs. If the user enters a response, the MS returns the response to the MSC.

8.1.3   Roamer initiated USSD operation

Figure 8-3 shows a roamer-initiated USSD operation and message flow at the B and D interfaces. When a roamer invokes a USSD request (e.g., a prepaid roamer initiates USSD callback) by keying in the appropriate code (containing the HPLMN service code), the USSD handler within the MS invokes the USSD request by sending a REGISTER message to the network. The REGISTER message contains a process unstructured SS request invoke component. The serving MSC, on receiving a USSD request containing an HPLMN service code, sets up a transaction to the VLR and forwards the request unchanged. The MSC will act in a transparent mode to any further requests/responses for this transaction, passing them between the MS and the VLR without taking any action.

When a VLR receives a USSD request containing an HPLMN service code, it sets up a transaction to the HLR and forwards the request unchanged. The VLR then acts in a transparent mode to any further requests/responses for this transaction. Passing them between the MSC and the HLR without taking any action. When the HLR receives the USSD

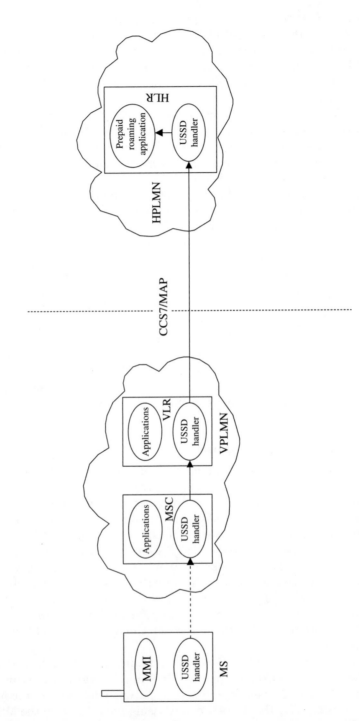

Figure 8-2 USSD handling at the GSM entities.

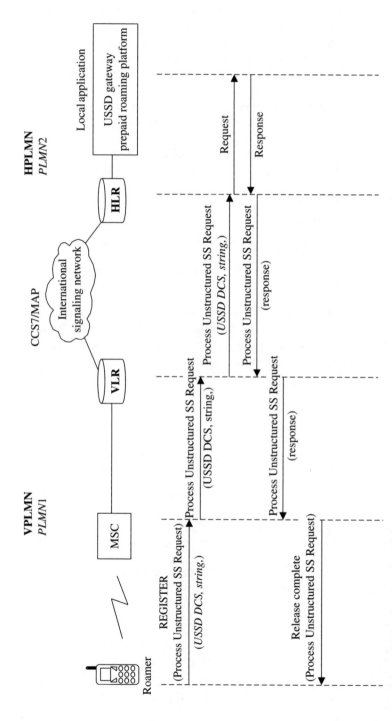

Figure 8-3 Roamer-initiated USSD operation.

request, it processes and invokes the appropriate application, i.e., a prepaid roaming platform. The application, as an option, may request further information in order to perform the requested operation or terminate the dialogue. If the application requests more information, it initiates a USSD request (see Section 8.1.4 for network-initiated requests), using the ongoing transaction. If the application decides to terminate the dialogue, the network side sends a release complete message. The MS can also terminate a dialogue by sending a release complete message.

Process unstructured SS request. The MAP process unstructured SS request procedure is used to relay USSD information between the:

- MSC and the VLR
- VLR and the HLR
- HLR and the gsmSCF (the gsmSCF is defined in the next section)

The process unstructured SS request message contains the following parameters:

USSD data coding scheme. This parameter contains the alphabet and the language information used for the unstructured information in a USSD operation. The coding of this parameter is according to the cell broadcast data coding scheme as specified in 3GPP TS 23.038.

Default alphabet: SMS default alphabet

Default language: Language unspecified

USSD string. This parameter contains a string of unstructured information. The string is sent either by the mobile user or the network. The contents of a string sent by the MS are interpreted as given in Table 8-1.

MSISDN. Originating subscriber international number. The MSISDN is an optional parameter.

The receiving entity, on unsuccessful outcome of the service, returns a user error. The possible error types are as follows:

- *System failure:* This indicates that the requested task could not be completed because of a problem in another entity.
- *Data missing:* This indicates that the context is missing in the received message.
- *Unexpected data value:* This error is returned if the receiver is not able to deal with the contents of the USSD string.
- *Call barred:* This indicates that the receiving entities cannot process the request because of barring of the initiated service.

- *Unknown alphabet:* This indicates that the receiving entity does not support the alphabet indicated in the USSD operation.

Figure 8-4 shows the protocol decodes of a process unstructured SS request message. The USSD string contains a service code, i.e., 138 in this case, and the MSISDN number of the called party.

8.1.4 Network initiated USSD operations

When a local application associated with a HPLMN HLR is to send a USSD request or notification to the MS, the HLR sets up a transaction to the VPLMN VLR where the subscriber is currently registered and sends the operation, i.e., MAP unstructured SS request or unstructured SS notify. The VLR, on receiving the operation, sets up a transaction to the MSC, where the subscriber is currently registered and passes the operation unchanged to it. The MSC, on receiving the request, sets up a transaction with the MS and invokes a USSD request by sending a REGISTER message to the MS. The register message contains the unstructured SS request or unstructured SS notify invoke component. The MS, on receiving the USSD request/notification, analyzes the data-coding scheme and decides whether the USSD operation is MMI mode or application mode.

The actions taken in the MMI mode are:

- For a USSD request, the MS displays the text provided and waits for the user inputs. If the user keys in the response, the MS sends it to the HLR/application via MSC, using the same transaction. If the user decides to release the transaction, the MS also releases the transaction.

- For USSD notify, the MS displays the text provided and sends back a response.

The actions taken in the application mode are:

- For an USSD request, the MS passes the message to the application in the ME, SIM/USIM, or TE and waits for the response. On receiving a response from the application, it passes it to the initiating entity via the MSC, using the same transaction. If the application decides to release the transaction, the MS also releases the transaction.

- For an USSD notify, the MS passes the message to the application and sends back a response.

Figure 8-5 shows a network-initiated USSD request. The same procedures are followed as in the case of USSD notify.

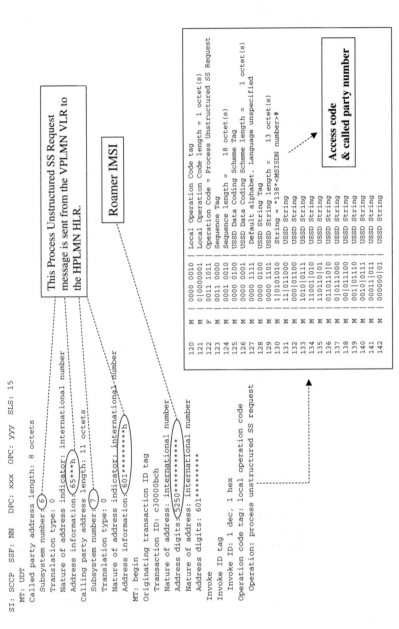

Figure 8-4 Process unstructured SS request message decode.

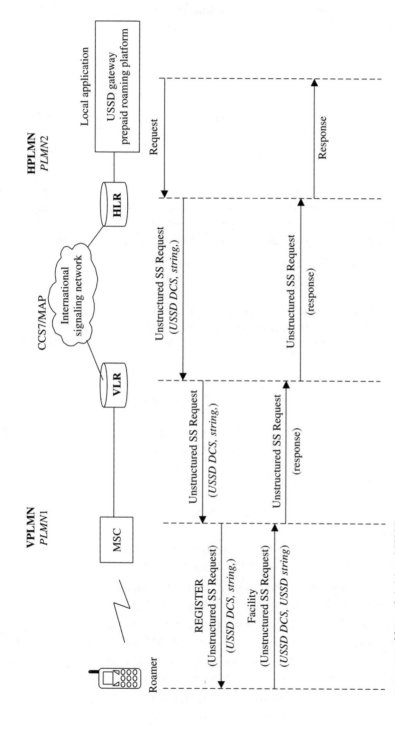

Figure 8-5 Network-initiated USSD request.

Unstructured SS request.   The MAP unstructured SS request procedure is used between the:

- gsmSCF and the HLR
- HLR and the VLR
- VLR and the MSC

This procedure is used when the invoking entity requires information from the roamer in connection with the USSD service handling.

The unstructured SS-Request message contains the following parameters:

USSD data coding scheme.   This parameter contains the alphabets and the language information used for the unstructured information in a USSD operation. The coding of this parameter is according to the cell broadcast data coding scheme as specified in 3GPP TS 23.038.

Default alphabet: SMS default alphabet

Default language: Language unspecified

USSD string.   This parameter contains a string of unstructured information. The string is sent either by the mobile user or by the network. The contents of a string sent by the MS are interpreted in Table 8-1.

Alerting pattern.   This parameter is an indication that can be used by the MSC to alert the user in a specific manner in the case of mobile terminating traffic (switched call or USSD). That indication can be an alerting level or an alerting category.

The responder, on unsuccessful outcome of the service, returns a user error. The possible error types are as follows.

- *System failure:* This indicates that the requested task could not be completed because of a problem in another entity.

- *Data missing:* This indicates that the context is missing in the received message.

- *Unexpected data value:* This error is returned if the receiver is not able to deal with the contents of the USSD string.

- *Unknown alphabet:* This indicates that the receiving entity does not support the alphabet indicated in the USSD operation.

- *Illegal subscriber:* The receiving entity indicates that the delivery of the USSD failed because the destination MS failed authentication.

- *Illegal equipment:* The receiving entity indicates that the delivery of the USSD failed because the destination MS failed the IMEI check.

- *USSD busy:* This indicates that the USSD handler at the receiving entity is busy and cannot handle further requests.

Unstructured SS notify. The MAP unstructured SS notify procedure is used to send a notification to a roamer in connection with the USSD service. It is used between the following network nodes.

- gsmSCF and the HLR
- HLR and the VLR
- VLR and the MSC

The unstructured SS notify message contains the following parameters:

USSD data coding scheme. This parameter contains the alphabet and the language information used for the unstructured information in a USSD operation. The coding of this parameter is according to the cell broadcast data coding scheme, as specified in 3GPP TS 23.038.

Default alphabet: SMS default alphabet

Default language: Language unspecified

USSD string. This parameter contains a string of unstructured information. The string is sent either by the mobile user or by the network. The contents of a string sent by the MS are interpreted as given in Table 8-1.

Alerting Pattern. This parameter is an indication that can be used by the MSC to alert the user in a specific manner in the case of mobile terminating traffic (switched call or USSD). That indication can be an alerting level or an alerting category.

Figure 8-6 shows a notification sent to the roamer, informing of the credit balance, using unstructured SS notify procedure.

The responder, on unsuccessful outcome of the service, returns a user error. The possible error types are as follows.

- *System failure:* This indicates that requested task could not be completed because of a problem in another entity.

- *Data missing:* This indicates that the context is missing in the received message.

- *Unexpected data value:* This error is returned if the receiver is not able to deal with the contents of the USSD string.

- *Unknown alphabet:* This indicates that the receiving entity does not support the alphabet indicated in the USSD operation.

- *Illegal subscriber:* The receiving entity indicates that the delivery of the USSD failed because the destination MS failed authentication.

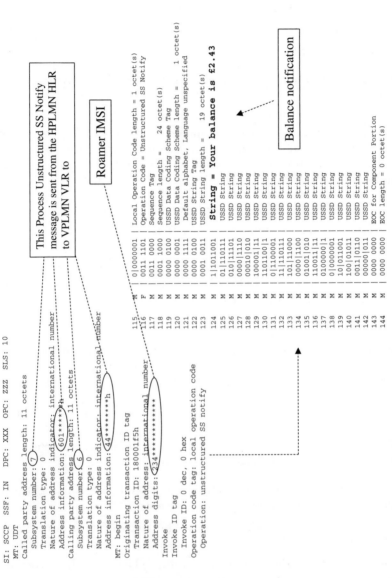

```
SI: SCCP  SSF: IN   DPC: XXX  OPC: ZZZ   SLS: 10
MT: UDT
  Called party address length: 11 octets
    Subsystem number: 7
    Translation type: 0
    Nature of address indicator: international number
    Address information: 601******h
  Calling party address length: 11 octets
    Subsystem number: 6
    Translation type: 0
    Nature of address indicator: international number
    Address information: 44********h
MT: begin
  Originating transaction ID tag
    Transaction ID: 18001f5h
    Nature of address: international number
    Address digits: 234************
  Invoke
    Invoke ID tag
    Invoke ID: 0 dec, 0 hex
    Operation code tag: local operation code
    Operation: unstructured SS notify
```

115	M	0|0000001	Local Operation Code length = 1 octet(s)
116	F	0011 1101	Operation Code = Unstructured SS Notify
117	M	0011 0000	Sequence Tag
118	M	0001 1000	Sequence length = 24 octet(s)
119	M	0000 0100	USSD Data Coding Scheme Tag
120	M	0000 0001	USSD Data Coding Scheme length = 1 octet(s)
121	M	0000 1111	Default alphabet, Language unspecified
122	M	0000 0100	USSD String Tag
123	M	0001 0011	USSD String length = 19 octet(s)
124	M	1|1011001	**String = Your balance is £2.43**
125	M	01|110111	USSD String
126	M	010|11101	USSD String
127	M	0000|1110	USSD String
128	M	00010|010	USSD String
129	M	100001|11	USSD String
130	M	1101100|1	USSD String
131	M	0|1100001	USSD String
132	M	11|110111	USSD String
133	M	011|11000	USSD String
134	M	0000|1100	USSD String
135	M	01001|010	USSD String
136	M	110011|11	USSD String
137	M	0100000|1	USSD String
138	M	0|0000001	USSD String
139	M	10|011001	USSD String
140	M	100|01011	USSD String
141	M	0011|0110	USSD String
142	M	00000|011	USSD String
143	M	0000 0000	EOC for Component Portion
144	M	0000 0000	EOC length = 0 octet(s)

This Process Unstructured SS Notify message is sent from the HPLMN HLR to VPLMN VLR to

Roamer IMSI

Balance notification

Figure 8-6 Unstructured SS notify request message decode.

- *Illegal equipment:* The receiving entity indicates that the delivery of the USSD failed because the destination MS failed the IMEI check.

- *USSD busy:* This indicates that the USSD handler at the receiving entities is busy and cannot handle further requests.

8.1.5   Prepaid roaming—USSD callback scenario

Figure 8-7 shows a typical callback sequence for prepaid roaming and the actions taken by each entity. The implementation of prepaid roaming application is vendor specific, and hence the call sequence shown here may not be valid in all cases.

1. A PLMN A subscriber (MSISDN A), currently roaming in PLMN B, keys in the USSD string (*111* C#) to initiate an outgoing call to subscriber-C, a fixed line subscriber that belongs to network C.

2. The USSD handler at the MS, on recognizing a valid USSD string, sends a **register** message with an invoke **process unstructured SS request** to the serving MSC. This message contains the USSD data coding scheme and USSD string.

3. The USSD handler within MSC analyzes the service code (see Table 8-1) and, on realising that the service code is not meant for its own applications, passes the USSD request to the VLR, using the MAP **process unstructured SS request** procedure.

4. As the service code 111 is reserved for the HPLMN, the VLR invokes the MAP **process unstructured SS request** procedure toward the HPLMN HLR. If the alphabet used for the message is understood by the HLR, then the message is fed to an application contained locally in the HLR, or to the gsmSCF, or to a secondary HLR where the USSD application is located. If the alphabet is not understood. then the error message unknown alphabet is returned.

5. A USSD acknowledgment is sent by the HLR to the roamer by the same transaction. (Further actions taken by the application are implementation specific and vendor dependent. Steps 6 to 11 as described here are just for explanation purposes.)

6. The application performs precall checks. For example, it checks if the caller is authorized to make such a call and if there is enough credit balance in the caller's account. If the caller does not qualify, an appropriate notification is sent by using the unstructured **SS notify** procedure. If the caller qualifies, the application initiates a mobile terminating call to the calling party first. The MAP **send routing information** (SRI) procedure is invoked toward the HLR to

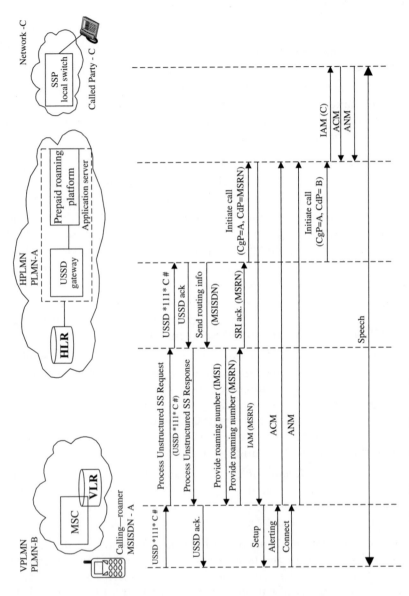

Figure 8-7 Prepaid roaming call scenario.

get the routing information necessary to establish the call. The HLR, on receiving the SRI message, invokes the **provide roaming number** (PRN) procedure toward the serving MSC/VLR to get the MSRN assigned to the roamer.

7. Once the MSRN is known, the application using the ISUP procedure initiates an outgoing call. (The application server implementation is vendor dependent. As a minimum, it should have call processing capabilities and should be connected to a MSC/STP by CCS7 ISUP links. It should also have the capability to send SMS notification to a roamer via SMSC.)

8. On answer from the roamer, a suitable announcement or tone is fed to the roamer indicating the call progress.

9. The prepaid roaming platform then initiates a second outgoing call toward called party C, using ISUP procedures.

10. On answer from called party C, a circuit-switched call is established between the roamer and the called party.

11. The application server monitors the call. On disconnection from any of the parties, it releases the call and frees up resources. In cases where the credit balance is exhausted during the call, an appropriate notification is sent to the roamer before call disconnection.

8.2   Prepaid Roaming Using CAMEL

Customized applications for mobile enhanced logic (CAMEL) were introduced by ETSI to incorporate intelligent network functionalities in GSM networks. The capabilities offered by CAMEL enable GSM operators to implement operator-specific services based on IN service logic that are seamlessly available to subscribers even if they are roaming in a foreign network.

CAMEL is built on the Intelligent Network Application Part (INAP), which is used in fixed line networks. INAP does not support mobility, and hence is not suitable for mobile networks. CAMEL also helps mobile operators to move onto a common standard, as many vendor-specific proprietary INAP versions are available in the market.

Unlike USSD, CAMEL is not a supplementary service but is a network feature. The partner networks can upgrade their networks to support CAMEL in order to enable seamless services. At this time, CAMEL is progressively being implemented in many networks. It has yet to achieve a global footprint.

CAMEL specifications were developed and released in phases to facilitate early adoption and implementation. Each new version of CAMEL is backward compatible.

The CAMEL Phase I specification was released in 1997, with capabilities limited to basic mobile-originated and mobile-terminated call activities.

The CAMEL Phase II specification was released in 1998. The new functionalities include support of intelligent peripherals, which provide capability to insert announcements and tones, in-band subscriber interaction using voice prompt, and information collection. Phase 2 also supports monitoring of invocation of selected supplementary services, USSD, and the charging function to enable prepaid calls.

CAMEL Phase III and IV, which were released subsequently, include specific functions for mobility management, GPRS, and UMTS.

8.2.1 CAMEL architecture

Figure 8-8 shows the CAMEL-based network architecture. This includes home network, visited network, and interrogating network.

The home network represents the HPLMN of a CAMEL subscriber or a CAMEL-enabled roamer. At HPLMN, the HLR and the gsmSCF are two functional entities involved in CAMEL procedures. The HLR stores CAMEL subscriber information (CSI) and transfers it to other entities that take part in enabling CAMEL services. The list of CSIs, their contents, and their functions are described in the next section. The GSM service control function (gsmSCF) is a new entity that consists of service logic for operator-specific services (OSSs). The HLR and the gsmSCF communicate by using the MAP protocol. The gsmSSF functionality, which resides in the VPLMN or interrogating network, communicates with gsmSCF by using the CAMEL application protocol (CAP). The HLR and the gsmSCF are addressed by other entities using global titles.

The PLMN where the CAMEL subscriber is currently roaming is called the visiting network, i.e., VPLMN. The VPLMN handles all the services invoked by the roamers. The VLR stores CAMEL subscription information received from an HLR during the update location procedure or at any other time as a result of changes in the roamer's data. The VLR also provides the roamer status if interrogated by other entities. The GSM service switching function (gsmSSF), which resides within MSC, acts as an interface between the MSC and the gsmSCF. The gsmSSF initiates the dialogue and gets instructions from the gsmSCF to handle a CAMEL service invoked by a roamer in its coverage area. The gsmSSF and gsmSCF communicate to each other by using the CAMEL application protocol (CAP).

The PLMN that interrogates the HPLMN for information to handle mobile terminating call is termed the interrogating network (IPLMN). The GMSC and the gsmSSF are the IPLMN entities, which interact with the HPLMN HLR and the gsmSCF to get further instructions on how to handle the terminating call. The MAP protocol is used on the

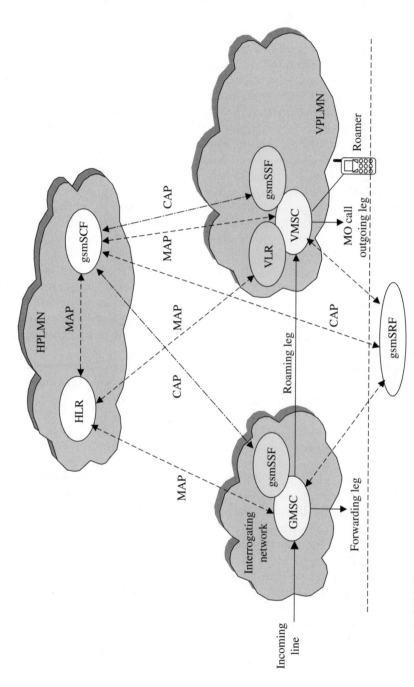

Figure 8-8 CAMEL architecture.

GMSC-HLR interface. The interface is used to exchange routing information, subscriber status, location information, subscription information, and suppression of announcements. In addition, CAMEL-related subscriber data that is passed to the IPLMN is sent over this interface.

The GSM specialized resource function (gsmSRF) is a functional entity that provides a pool of specialized resources such as tones and announcements. For example, if the prepaid roamer is in an active call and the credit balance is running out, then gsmSRF can be used under command of gsmSCF to play a warning announcement indicating low balance. The gsmSRF component was added in CAMEL Phase 2, and it can reside in the home, visited, or interrogating network. The gsmSCF uses the CAP protocol to communicate with the gsmSRF.

The HPLMN, VPLMN, and interrogating network must support CAMEL to provide operator-specific services (OSSs) to a roamer.

Figure 8-9 shows the functional architecture to support GPRS interworking for CAMEL. The architecture is applicable to the third phase of CAMEL or higher.

The HLR stores additional CSI for the GPRS, i.e., GPRS-CSI, and passes it to the SGSN, using the HLR-SGSN interface. The gprsSSF at the SGSN sends a request for the instructions to the gsmSCF when a CAMEL-enabled roamer invokes the services. The gsmSCF then communicates with gsmSSF to control a GPRS session or an individual PDP context.

The Figure 8-10 shows the functional architecture to support mobile-originating short message service (MO-SMS) and mobile-terminating short message service (MT-SMS) for CAMEL.

The HLR stores MO-SMS-CSI or MT-SMS-CSI or both, depending on the support required. The MO-SMS-CSI contains subscription information for subscribers that require CAMEL support of MO-SMS. The MT-SMS-CSI contains subscription information for subscribers that require CAMEL support of MT-SMS. One or both of MO-SMS-CSI and MT-SMS-CSI are transferred to the VLR or to the SGSN on location update and restore data or when MO-SMS-CSI or MT-SMS-CSI has changed. The VPLMN MSC/VLR examines MO-SMS-CSI and MT-SMS-CSI to determine the invocation of service logic on MO-SMS submission or MT-SMS delivery.

In cases where the GPRS bearer is used for SMS, the SGSN determines if a service logic shall be invoked by examining MO-SMS-CSI and MT-SMS-CSI received from the HLR.

Figure 8-11 shows the functional architecture needed to support CAMEL handling of USSD to/from gsmSCF. USSD CAMEL subscription information (U-CSI) and USSD global subscriber information (UG-CSI) is stored in the HLR to support USSD for CAMEL. Unlike other CSIs,

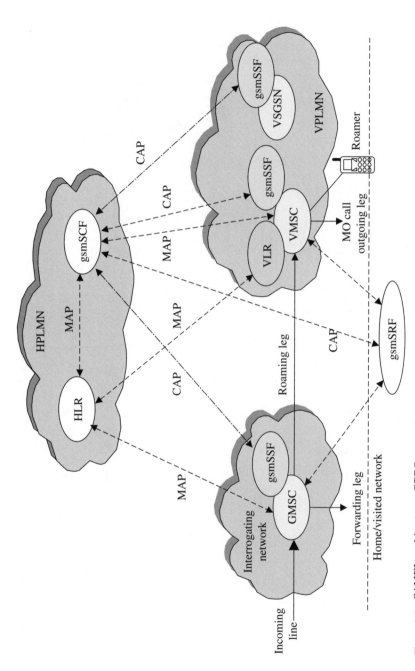

Figure 8-9 CAMEL architecture: GPRS support.

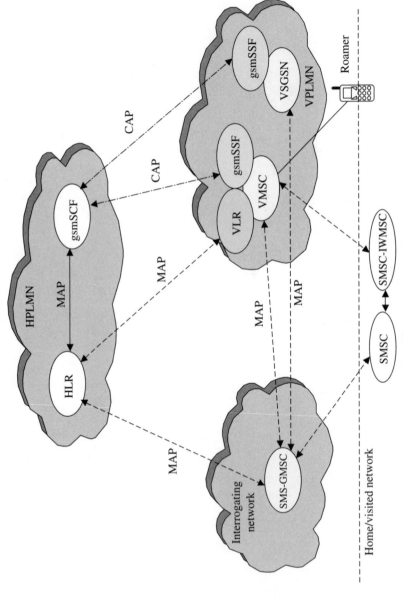

Figure 8-10 Support for CAMEL control of MSC/SGSN switched MO/MT-SMS.

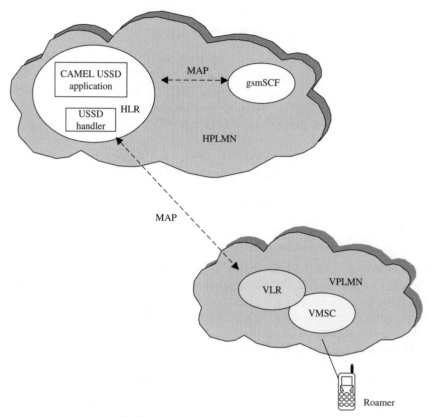

Figure 8-11 Handling of USSD to/from a CAMEL subscriber.

these are never transferred to any other entities. U-CSI contains the subscriber-specific information, while UG-SCI contains global data applicable to all subscribers.

The new-functionality CAMEL USSD application is added in the HLR to support USSD for CAMEL. The actions of the HLR on receiving USSD are as follows:

1. On receiving a USSD initiated by a roamer in the visited network, the USSD handler examines the USSD service code. If the service code does not match with the service codes for which a USSD application within HLR needs to be invoked, the USSD handler routes the USSD to the CAMEL USSD application.

2. On receiving a USSD initiated by the gsmSCF, the HLR establishes the transaction with the VLR/MSC where the roamer is currently roaming

and transfers the information unchanged. The further information flow between the gsmSCF and the VMSC happens transparently.

The CAMEL USSD application checks the U-CSI data assigned to the specific subscriber. If the service code in the USSD matches with the service code present in the U-CSI, the USSD is routed to the gsmSCF. If the service code does not match, that means the subscriber does not have U-CSI defined, and then the CAMEL USSD application checks the UG-CSI data assigned to the HLR. If the service code is present in the UG-CSI, then the USSD is routed to the gsmSCF given by the gsmSCF address stored against the service code in the UG-CSI. If the service code is not present in U-CSI or UG-CSI, an error (unknown application) is returned to the USSD handler.

8.2.2 Points in call and detection points

The various events such as call origination, answer, and disconnection during establishing, maintaining, or tearing down a call are referred as the points in call (PICs). The PICs provide a view of a state or an event in which call processing logic within the MSC/VLR/SGSN may initiate an appropriate action.

Detection points (DPs) represent transitional events that occur between PICs. The gsmSSF makes these DPs visible to the gsmSCF as and when they are encountered. This allows gsmSCF to control subsequent handling of the call.

The service logic that is loaded in the network nodes to carry out the call processing on encountering the detection points is termed a trigger.

The DP can be armed or disarmed. "Arming" a DP means that the gsmSSF must notify the gsmSCF on encountering it. If the DP is not armed, the processing node continues processing of a call on its own, without involvement of the gsmSCF. The DPs can be armed statically or dynamically.

The CAMEL specifications identify three different types of DP:

- *Trigger detection point–request (TDP-R):* The DP is statically armed. The processing is suspended when the TDP is encountered.

- *Event detection point–request (EDP-R):* The DP is dynamically armed within the context of a CAMEL-controlled relationship. The processing is suspended on encountering the EDP. The gsmSSF then processes the call according to the instructions from the gsmSCF.

- *Event detection point–notification (EDP-N):* Like EDP-R, this DP is also dynamically armed within the context of a CAMEL-controlled relationship. However, the processing of the call continues without suspension on encountering an EDP-N

8.2.3   CAMEL subscriber information

The HLR stores the following CAMEL subscription information (CSI): (1) that CAMEL support is required for the subscriber and (2) the identities of the CAMEL-specific entities (CSEs) to be used for that support.

Circuit-switched call CSIs.   Table 8-2 lists CSIs relevant to circuit-switched calls.

TABLE 8-2   Circuit-Switched Call CSIs

CAMEL subscription information (CSI)	Description
O-CSI *Originating CSI*	The O-CSI identifies the roamer as having originated CAMEL services. It contains the trigger information that is required to invoke CAMEL service logic for the roamer-originated call. The HPLMN HLR sends O-CSI to the VPLMN VLR/MSC in the insert subscriber data operation during the location update procedure or whenever it gets changed, e.g., because of administrator action. It is also transferred to the interrogating network when a GMSC receives an incoming call for a subscriber roaming in a visited network and interrogates the HLR for routing information, using the send routing information procedure.
D-CSI *Dialed service CSI*	The D-CSI contains trigger information that is required to invoke CAMEL service logic for subscriber-dialed services. The HPLMN HLR sends D-CSI to the VPLMN VLR/MSC in the insert subscriber data operation as part of the location update procedure or whenever it gets changed, e.g., because of administrator action. It is also transferred to the interrogating network when a GMSC receives an incoming call for a subscriber roaming in a visited network and interrogates HLR for routing information using the send routing information procedure.
T-CSI *Terminating CSI*	The T-CSI identifies the roamer as having terminated CAMEL services. It contains the trigger information that is required to invoke CAMEL service logic for a mobile-terminating call in the GMSC.
VT-CSI *Visited-MSC-terminating CSI*	The VT-CSI contains trigger information that is required to invoke CAMEL service logic for mobile-terminating calls in the VMSC. The HPLMN HLR sends VT-CSI to the VPLMN VLR/MSC in the insert subscriber data operation as part of the location update procedure.
TIF-CSI *Translation information flag CSI*	The TIF-CSI is used in the HLR for registering short forwarded-to Numbers (FTNs). When TIF-CSI is present, the subscriber is allowed to register short FTNs. When the subscriber invokes call deflection, the TIF-CSI in the VPLMN allows the subscriber to deflect to short deflected-to numbers. The TIF-CSI is transferred to the VPLMN in the insert subscriber data.
N-CSI *Network CSI*	The N-CSI identifies services offered on a per-network basis by the serving PLMN operator for all subscribers. This CSI is stored in the MSC.

Each CSI contains a set of information elements. Table 8-3 lists information elements for each CSI.

TDP list. The TDP list indicates on which detection point triggering shall take place. The list of all detection points on which triggering takes place is as follows:

For O-CSI, the detection points are:

- DP collected info
- DP route select failure

For T-CSI, the detection points are:

- DP terminating attempt authorized
- DP T busy
- DP T no answer

For VT-CSI, the detection points are:

- DP terminating attempt authorized
- DP T busy
- DP T no answer

gsmSCF address. The gsmSCF address indicates the address to be used to access the gsmSCF for a particular subscriber. The address is based on the E.164 addressing scheme and is used for routing. There may be more than one gsmSCF associated with a TDP.

Service key. The service key identifies to the gsmSCF the service logic to be used. A service key is associated with each DP criterion.

DP criteria. The DP criteria indicate if the gsmSSF needs to request instructions from the gsmSCF.

TABLE 8-3   Circuit-Switched Call CSI Contents

Contents	O-CSI	D-CSI	T-CSI	VT-CSI	TIF-CSI	N-CSI
TDP list	X		X	X	X	
gsmSCF address	X	X	X	X		
Service key	X	X	X	X		
DP criteria	X	X	X	X		
Default call handling	X	X	X	X		
CAMEL capability handling	X	X	X	X		
CSI state	X	X	X	X		
Notification flag	X	X	X	X	X	
Translation information flag					X	
List of services						X

Default call handling. The default call handling element indicates whether the call shall be released or continued as requested if there is an error in the dialogue between gsmSSF and gsmSCF or if the call is subject to call gapping in the gsmSSF. A default call handling element is associated with each service key.

CAMEL capability handling. The CAMEL capability handling element indicates the phase of CAMEL that is requested by the gsmSCF for the service. The HLR does not include in a CSI that it sends to a GMSC any data for a CAMEL phase later than that which the CAMEL capability handling indicates. The CAMEL capability handling may be different for the different CSIs. For example, O-CSI may have a CAMEL capability handling value to indicate CAMEL Phase 3, while T-CSI may contain a value indicating CAMEL Phase 2 support.

To enable interworking between networks supporting different phases of CAMEL, the HLR decides on a subscriber basis to apply operator-determined barring, perform normal call handling, or perform operator-specific handling.

CSI state. This indicates whether the CSI (i.e., O-CSI/D-CSI/T-CSI/VT-CSI) is active or not.

Notification flag. The notification flag indicates whether a change in the CSI shall trigger a notification or not.

Translation information flag. The TIF-CSI in the CAMEL subscriber data indicates that:

■ When the subscriber registers a forwarded-to number supplementary service, the HLR shall not attempt to perform any translation, number format checks, prohibited FTN checks, or call barring checks.

■ When the subscriber invokes the call deflection supplementary service, the VLR shall not attempt to perform any translation, number format checks, prohibited DTN checks, or call barring checks.

List of Services. The list of services in N-CSI identifies services offered on a per-network basis by the serving PLMN operator for all subscribers.

USSD CSIs. Table 8-4 lists CSIs to support USSD services. The contents of USSD CSIs are:

Service code. The service code identifies a specific application in a gsmSCF that interacts with the user via the USSD.

gsmSCF address. The gsmSCF address indicates the address to be used to access the gsmSCF for a particular service. The address is based of on the E.164 addressing scheme and is used for routing.

TABLE 8-4   USSD-CSI

CAMEL subscription information (CSI)	Description
U-CSI *USSD CSI*	The U-CSI contains trigger information that is used to invoke a USSD application in the CAMEL service environment for the *served* subscriber. The U-CSI is held in the HLR and never sent to any other node. The U-CSI was introduced in CAMEL Phase 2.
UG-CSI *USSD general CSI*	The UG-CSI contains trigger information that is used to invoke a USSD application in the CAMEL service environment for *all* subscribers. The UG-CSI is held in the HLR and never sent to any other node. The UG-CSI was introduced in CAMEL Phase 2.

SS notification CSI.   Table 8-5 lists the supplementary services notification CSI and its functions.

The SS notification CSI contents are:

Notification criteria.   This data indicates for which supplementary services notifications shall be sent. The supplementary services that may be indicated are ECT, CD, CCBS, and MPTY.

gsmSCF address.   This is the E.164 address of a gsmSCF. It is used for routing purposes to access the gsmSCF for a particular subscriber.

CSI state.   The CSI state indicates whether the SS-CSI is active or not.

Notification flag.   The notification flag indicates whether the change of the SS-CSI shall trigger a notification on change of subscriber data or not.

GPRS CSI.   Table 8-6 lists the CSI to support GPRS.

The contents of the GPRS CSI are:

gsmSCF address.   This is the address of the gsmSCF in ITU-T E.164 format. It is used for routing purposes to access the gsmSCF for a particular subscriber.

TABLE 8-5   SS Notification CSIs

CAMEL subscription information (CSI)	Description
SS-CSI *Supplementary services invocation notification CSI*	The SS-CSI is used to notify the CSE about the invocation of certain supplementary services at the VLR/MSC. The HPLMN HLR sends SS-CSI to the VMSC in the insert subscriber data element as part of the location update procedure or whenever it gets changed, e.g., because of administrator action. The SS-CSI was introduced in CAMEL Phase 2.

TABLE 8-6   GPRS-CSI

CAMEL subscription information (CSI)	Description
GPRS CSI *GPRS CAMEL subscription information*	The GPRS-CSI contains trigger information that is required to invoke CAMEL service logic for GPRS sessions and PDP contexts. The HPLMN HLR sends GPRS-CSI to the SGSN during GPRS attach and inter SGSN routing area update procedure or whenever it gets changed, e.g., because of administrator action. The GPRS-CSI was introduced in CAMEL Phase 3.

Service key. The service key identifies to the gsmSCF the service logic to be used.

Default GPRS handling. The default GPRS handling indicates if the GPRS session or PDP context shall be released or continued as requested in case of error in the gprsSSF to gsmSCF dialogue.

TDP list. The TDP list indicates the detection point on which triggering shall take place.

CAMEL capability handling. To enable interworking, CAMEL capability handling indicates the phase of CAMEL that gsmSCF requests for the service.

CSI state. The CSI state indicates whether the GPRS-CSI is active or not.

Notification flag. The notification flag indicates whether the change of the GPRS-CSI shall trigger a notification on change of subscriber data or not.

SMS CSIs. Table 8-7 lists CSIs to support SMS. The contents of the MO-SMS CSI are:

gsmSCF address. This is the E.164 address of a gsmSCF. It is used for routing purposes to access the gsmSCF for a particular subscriber.

Service key. The service key identifies to the gsmSCF the service logic to be used.

Default SMS handling. The default SMS handling element indicates whether the SMS submission/short message delivery shall be released or continued as requested in case of error in the dialogue between gprsSSF and gsmSCF/gprsSCF.

TDP list. The TDP list indicates on which detection point triggering shall take place.

TABLE 8-7   SMS CSIs

CAMEL subscription information (CSI)	Description
MO-SMS CSI *Mobile-originated short message service CSI*	The HPLMN HLR sends MO-SMS-CSI to the VMSC in the insert subscriber data element as part of the location update procedure or whenever it gets changed, e.g., because of administrator action. The MO-SMS-CSI was introduced in CAMEL Phase 3.
MT-SMS CSI *Mobile-terminated short message service CSI*	The HPLMN HLR sends MT-SMS-CSI to the VMSC in the insert subscriber data element as part of the location update procedure or whenever it gets changed, e.g., because of administrator action. The MT-SMS-CSI was introduced in CAMEL Phase 3.

- For MO-SMS-CSI, the only defined DP is SMS collected info.
- For MT-SMS-CSI, the only defined DP is SMS delivery request.

DP criteria.   The DP criteria indicate whether the SMS SSF shall request the gsmSCF for instructions. This information element is present only in case of MT-SMS-CSI.

CAMEL capability handling.   The CAMEL capability handling indicates the phase of CAMEL that is asked by the gsmSCF for the service. This parameter is set to CAMEL Phase 3 for MO-SMS-CSI and CAMEL Phase 4 for MT-SMS-CSI.

CSI state.   The CSI state indicates whether the MO-SMS-CSI/MT-SMS-CSI is active or not.

Notification flag.   The notification flag indicates whether the change of the MO-SMS-CSI/MT-SMS-CSI shall trigger notification on change of subscriber data.

Mobility management-CSI.   Table 8-8 lists the CSIs to support mobility management.

The contents of the mobility management CSI are:

Mobility management triggers.   This data indicates which mobility management events shall result in a notification to the gsmSCF. One or more events may be marked per subscriber. These events are:

- Location update in the same VLR service area.
- Location update to another VLR service area.
- IMSI attach.

TABLE 8-8   Mobility Management CSIs

CAMEL subscription information (CSI)	Description
M-CSI *Mobility management CSI*	M-CSI is used to notify the CSE about mobility management events for the CAMEL subscriber. The HPLMN HLR sends M-CSI to the VMSC in the insert subscriber data element as part of the location update procedure or whenever it gets changed, e.g., because of administrator action.
MG-CSI *Mobility management GPRS CSI*	The MG-CSI is used to notify the CSE about mobility management events for the GPRS subscriber. The HPLMN HLR sends MG-CSI during the routing area update, or restore data procedure, or whenever it gets changed, e.g., because of administrator action.

- MS-initiated IMSI detach (explicit detach) and network-initiated IMSI detach (implicit detach).

gsmSCF address.   This is the E.164 address of the gsmSCF to which the mobility management event notification needs to be sent. It is used for routing purposes to access the gsmSCF for a particular subscriber.

Service key.   The service key identifies to the gsmSCF the service logic to be used.

CSI state.   The CSI state indicates whether the M-CSI is active or not.

Notification flag.   The notification flag indicates whether the change of the M-CSI shall trigger a notification on change of subscriber data or not.

8.2.4   Basic call state model

Basic call state model (BCSM) is used to describe the actions and different phases of the call processing in the MSC/GMSC/VMSC. The BCSM consists of two sets of call processing logics, i.e., originating BCSM (O-BCSM) and terminating BCSM (T-BCSM). The DP and points in call are two main components of a BCSM. Figure 8-12 illustrates the BCSM components.

O-BCSM.   O-BCSM is used to describe the actions taken by the serving MSC to handle a call invoked by a subscriber with active O-CSI. This can also be used to describe actions of the MSC/GMSC for the forwarded calls. Figure 8-13 shows PIC and DP at different phases of a call.

On receiving a setup message from an MS, if the serving MSC/VMSC finds an active CAMEL subscription information (CSI) in its VLR, the MSC

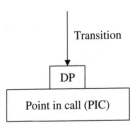

Figure 8-12   BCSM components.

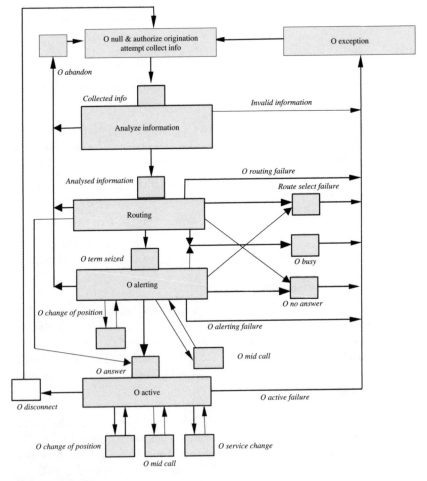

Figure 8-13   O-BCSM.

suspends the call and initiates the dialogue with the gsmSCF in the subscriber's HPLMN. This dialogue contains an initial DP message indicating that a statically armed DP2, i.e., collected info, has been encountered. The gsmSCF requests the serving MSC to monitor the detection points DP7, O answer, and DP9, O disconnect. The gsmSCF then instructs the MSC to continue call processing. In the case of a successful call scenario, when terminating party answers the call, the MSC sends a notification, i.e., DP7, O answer. The MSC continues to process and monitor the call under instructions from the gsmSCF. When either calling or called party terminates the call, the MSC reports back to the gsmSCF, indicating that DP9, O disconnect, has been encountered.

The detection points for the originating BCSM are described in Table 8-9

TABLE 8-9   O-BCSM DPs

CAMEL DP	Description
Collected info	This DP indicates that O CSI is active and has been analyzed. The dialed number has been received in the setup message sent by the initiating subscriber but is not yet analyzed.
Analyzed information	Indicates that the routing address and the nature of address is available and analyzed.
Route select failure	Indicates that call establishment failed because of the failure to select the route for the call.
O busy	A busy indication is received by the terminating party, i.e., ISUP REL message has been received by the serving MSC with the release cause code "busy."
O no answer	Indicates that one of the following events has occurred. ■ Serving MSC has received ISUP REL message with release cause code "no answer." ■ An application timer associated with O no answer has been triggered.
O term seized	An alerting indication is received from the terminating party, i.e., ISUP ACM has been received by the serving MSC.
O answer	Indication that the call is accepted and answered by the terminating party, i.e., an ISUP ANM message has been received by the serving MSC.
O midcall	Indication that a service feature request is received from the calling party via in-band DTMF signaling.
O change of position	Indication that the originating party has changed position.
O disconnect	Indication that either a disconnect message has been received from the originating party or a ISUP REL message with release cause code "normal release" has been received from the terminating side.
O abandon	Indication that a disconnect message has been received from the originating party during the call establishment phase.

T-BCSM.   In the case of mobile-terminating call, the IPLMN GMSC examines the called party MSISDN to identify its home PLMN. The GMSC then sends a request to the HPLMN HLR to get routing information, using the send routing information procedure. The HLR responds with the subscriber routing information, including the CSI of the called party. The GMSC checks the received CSI and acts accordingly. If the T-CSI is active and the trigger criterion of a DP is met, the call processing is suspended in order to get further instructions from the gsmSCF. An initial DP message is sent, informing the gsmSCF that a statically armed DP12 message, i.e., terminating attempt authorized was encountered in the T-BCSM. The gsmSCF requests the GMSC to monitor the DP's T answer and T disconnect. The gsmSCF then asks GMSC to continue the call processing. Once the called party answers, the GMSC sends the notification to the gsmSCF on DP T answer.

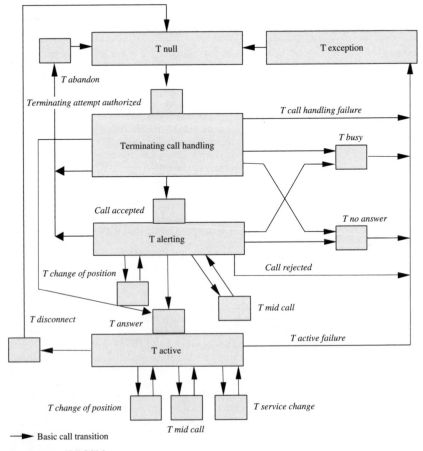

Figure 8-14   T-BCSM.

The call processing continues in accordance with gsmSCF, and the GMSC continues monitoring state call and notifies the gsmSCF on call disconnection by the calling or the called party.

Figure 8.14 describes possible PICs and DPs at different phases of a call.

The detection points for T-BCSM are described in Table 8-10.

8.2.5 CAMEL information flow

The following tables describe the information flow between CAMEL nodes to support circuit-switched calls, USSD, SMS, supplementary service invocation, and mobility management.

CAMEL operation for circuit-switched calls. Table 8-11 lists the operations in a CAMEL circuit-switched call.

CAMEL operation for USSD. Table 8-12 lists the CAMAL IF operations for USSD.

CAMEL operation for GPRS interworking. Table 8-13 lists the CAMEL operations for GPRS interworking.

CAMEL operation for short message services. Table 8-14 lists the CAMEL information flow for short message service.

TABLE 8-10 T-BCSM DPs

CAMEL DP	Description
Terminating attempt authorized	Indicates that the T-CSI/VT-CSI is active and analyzed.
T busy	Indicates that a busy indication is received from the destination exchange.
T no answer	This indicates that an application timer associated with the T no answer DP has expired.
Call accepted	This indicates that the called party is alerted.
T answer	This indicates that the call is answered by the terminating party.
T midcall	Indication that a service feature request is received from the terminating party via in-band DTMF signaling.
T change of position	Indicates that the terminating party has changed its position.
T disconnect	A disconnect indication is received from the originating or the terminating party.
T abandon	Indication that a disconnect message is received from the originating party during the call establishment phase.

TABLE 8-11   CAMEL Information Flow for a Circuit-Switched Call

CAMEL operation / information flow (IF)	Description
gsmSSF → gsmSCF	
Activity test ack	Response of IF activity test.
Apply charging report	A report in response to the apply charging request.
Call information report	Specific call information in response to the call information request.
Disconnect leg ack	Successful response to the disconnect leg IF.
Entity released	Sent to the gsmSCF to inform it of the release of a logical entity caused by exceptions/errors.
Event report BCSM	Used to notify about a call-related event once it has occurred—for example, events like answer and disconnect.
Initiate call attempt ack	The successful response to the initiate call attempt.
Initial DP	The gsmSSF uses this IF to request instructions from the gsmSCF when a trigger is detected at a DP in BCSM.
Move leg ack	The successful response to the move leg request.
Split leg ack	The successful response to the split leg IF.
gsmSCF → gsmSSF	
Activity test	Used to check if the gsmSCF and the gsmSSF relationship is in existence. Appropriate actions are taken to recover the relationship if no response is received.
Apply charging	Used to instruct the gsmSSF to apply charging mechanisms to control the call duration.
Call gap	Used to control the number of service requests sent to a gsmSCF
Call information request	This is used to instruct the gsmSSF to record specific information about a single call party and report back.
Cancel	Request to cancel all event DPs and report.
Connect	Request to the gsmSSF to perform call processing actions required to route a call to the destination.
Connect to resource	Request to connect a call to a specialized resource, i.e., gsmSRF.
Continue	Request to the gsmSSF to continue a previously suspended call.
Continue with argument	Request to the gsmSSF to continue a previously suspended call but with modified call setup information.
Disconnect forward connection	Request to disconnect a connection with the gsmSRF.
Disconnect forward connection with argument	Request to disconnect a connection with the gsmSRF. This is used to explicitly disconnect a connection that was previously established with a connect to resource or an establish temporary connection message.
Disconnect leg	Request to the gsmSSF to release a specific leg of a call while retaining all other legs of that call.
Establish temporary connection	Request to create a connection between an initiating gsmSSF and an assisting gsmSSF as a part of the assist procedure. It is also used to establish a connection between a gsmSSF and a gsmSRF.
Furnish charging information	Request to the gsmSSF to include call-related information in the CAMEL-specific logical call record.
Initiate call attempt	Request to the gsmSSF to establish a new call or add a new party to an existing established call.

TABLE 8-11   CAMEL Information Flow for a Circuit-Switched Call (Continued)

CAMEL operation / information flow (IF)	Description
gsmSCF → gsmSSF	
Move leg	Request to the gsmSSF to move a call leg to another call segment.
Play tone	Request to play a sequence of tones to a particular leg or call segment.
Release call	Request to tear down a call.
Request report BCSM event	Request to the gsmSSF to monitor for a call-related event and then send notification back on event detection.
Reset timer	Request to reset a specific timer.
Send charging information	Charging information to be sent to MS.
Split leg	Request to move a call leg from one call segment to another existing call segment or newly created call segment.
gsmSCF → gsmSRF(optional)	
Activity test	Used to check if the gsmSCF and the gsmSRF relationship is in existence. Appropriate actions are taken to recover the relationship if no response is received.
Cancel	Request to the gsmSRF to cancel a correlated previous information flow.
Play announcement	A request to play a single announcement/tone or sequence of announcements for the in-band interaction.
Prompt and collect user information	Used to interact with the calling party by announcement/ tone and to collect information such as digits or speech.
gsmSRF → gsmSCF	
Activity test ack	Response to the IF activity test.
Assist request instructions	Sent by the gsmSRF or by the assisting gsmSSF to associate the assist request with the initial DP.
Prompt and collect user information ack	Response to the IF prompt and collect user information.
Specialized resource report	Response to the request for resources by the gsmSCF in a play announcement message or a prompt and collect user information message.
gsmSCF → assisting SSF	
Activity test	This is used to check if the gsmSCF and the assist SSF relationship is in existence. Appropriate actions are taken to recover the relationship if no response is received.
Cancel	Request to cancel a correlated information flow.
Connect to resource	Request to connect a call to a specialized resource, i.e., gsmSRF.
Disconnect forward connection	Used to disconnect a previously established connection using a connect to resource message.
Play announcement	A request to play a single announcement/tone or sequence of announcements for the in-band interaction.
Prompt and collect user information	Used to interact with the calling party by announcement/ tone and to collect information such as digits or speech.
Rest timer	A request to reset a specific timer.

(Continued)

TABLE 8-11   CAMEL Information Flow for a Circuit-Switched Call *(Continued)*

CAMEL operation / information flow (IF)	Description
	assisting SSF → gsmSCF
Activity test ack	A response to IF activity test.
Assist request instructions	This is sent by the gsmSRF or by the assisting gsmSSF to associate the assist request message with the initial DP.
Prompt and collect user information ack	A response to the IF prompt and collect user information.
Specialized resource report	Response to the request for resources by the gsmSCF in a play announcement message or a prompt and collect user information message.
	HLR → VLR
Delete subscriber data	Used by the HLR to delete CAMEL subscription data from a VLR.
Insert subscriber data	The HLR uses this IF to update the VLR with subscriber specific data. CAMEL-specific subscriber information such as O-CSI and T-CSI are transferred by this procedure.
Provide subscriber info	The HLR can request anytime information on a subscriber currently roaming in a VPLMN. The VLR sends a response with roamer information such as subscriber location, IMEI, and SW version.
Provide roaming number	Used by the HLR to get the roaming number assigned to a roamer in a visited network.
	VLR → HLR
Insert subscriber data ack	Response to the IF insert subscriber data.
Provide subscriber info ack	Response to the IF provide subscriber info.
Update location	Used by the VLR to provide information about supported CAMEL phases to the HLR.
Restore data	Used by the VLR to provide information about supported CAMEL phases to the HLR.
	HLR → GMSC
Send routing info ack	Response to the IF send routing info. The HLR transfers routing information to enable the GMSC to route call to the destination MSC.
	GMSC → HLR
Send routing info	GMSC uses this operation to request information from the HLR to route an MT call.
	VMSC → GMSC
Resume call handling	Request to the GMSC to take over handling the call.

TABLE 8-11   CAMEL Information Flow for a Circuit-Switched Call *(Continued)*

CAMEL operation / information flow (IF)	Description
MSC → VLR	
Send info for ICA	Used to request the VLR to provide information to handle an outgoing call leg created by the gsmSCF.
Send info for incoming call	Used to request the VLR to provide information to handle an incoming call.
Send info for MT reconnected call	The MSC uses this to request the VLR to provide information to handle a reconnected MT call.
Send info for outgoing call	Used to request the VLR to provide information to handle an outgoing call.
Send info for reconnected call	The MSC uses this to request the VLR to provide information to handle a reconnected MO call.
VLR → MSC	
Complete call	Request to MSC to continue the connection of a call.
Continue CAMEL handling	Request to MSC to continue CAMEL-specific handling.
Process call waiting	Request to MSC to continue the connection of a waiting call.
Send info for ICA negative response	Negative response to indicate that the outgoing call leg for which the MSC requested subscription information will not be connected, given for a reason such as bearer service not allowed or call barred.
Send info for incoming call ack	Used to indicate that the incoming call for which the MSC requested subscription information should be forwarded.
Send info for incoming call negative response	Used to indicate that the incoming call for which the MSC requested subscription information should not be connected.
Send info for MT reconnected call ack	Used to instruct the MSC to continue the connection of a reconnected MT call.
Send info for MT reconnected call negative response	Used to indicate that the reconnected MT call for which the MSC requested subscription information should not be connected.
Send info for reconnected call ack	Request to MSC to continue the connection of a reconnected MO call.
Send info for reconnected call negative Ack	Used to indicate that the reconnected call for which the MSC requested subscription information should not be connected.
Internal MSC	
Perform Call forwarding ack	This IF is used to inform the MSC that the call forwarding is taking place.
gsmSCF → HLR	
Send routing info	A request for information from the HLR to route a gsmSCF-initiated call.
HLR → gsmSCF	
Send routing info ack	Response for IF send routing info.

TABLE 8-12   CAMEL Information Flow for USSD

CAMEL operation	Description
gsmSCF → HLR	
Unstructured SS request	Used to request data from the MS via the HLR.
Unstructured SS notify	Used to send data to the MS via the HLR.
Process unstructured SS data ack	Used to send a response to the MS via the HLR for the MS-initiated request.
HLR → gsmSCF	
Unstructured SS request ack	MS response to the gsmSCF-initiated IF unstructured SS request. The response is sent to the gsmSCF via the HLR.
Unstructured SS notify ack	Used by the MS to acknowledge to the gsmSCF the receipt of notification.
Process unstructured SS data	The MS requests data from the gsmSCF via the HLR.
Process unstructured SS request	The MS requests data from the gsmSCF via the HLR.
Begin subscriber activity	Used by the HLR to start subscriber activity toward the gsmSCF for USSD.

CAMEL operation for SS notification. Table 8-15 lists the CAMEL information flow for SS notification.

CAMEL operation for mobility management. Table 8-16 lists the CAMEL information flow for mobility management.

8.2.6   Prepaid roaming-CAMEL call scenario

Figure 8.15 shows the signaling message flow for an outgoing call initiated by a roamer in a visited network. For example purposes, MSISDN-A, which belongs to the PLMN-A, is shown visiting in the PLMN-B. Assume that the roamer is already authenticated and registered in the PLMN-B.

1. The roamer dials in the destination number that belongs to the PLMN-C.

2. The serving VMSC receives a setup message and checks subscription data for A in its VLR to determine if it has active O-CSI. The VLR has updated subscription data received previously during the update location procedure with the HPLMN. As explained in Section 8.2.3, the O-CSI identifies the roamer as having originating

TABLE 8-13   CAMEL Information Flow for GPRS Interworking

CAMEL operation	Description
	gsmSCF → gsmSCF
Activity test GPRS ack	Response to IF activity test.
Apply charging report GPRS	Report sent by the gprsSCF to the gsmSSF in response to the IF applying charging GPRS.
Entity released GPRS	The gprsSCF informs the gsmSCF that a GPRS session has been detached.
Event report GPRS	A GPRS event notification sent to the gsmSCF in response to the IF request report GPRS event.
Initial DP GPRS	Used by the gprsSCF to get instructions from the gsmSCF on detection of a trigger at a detection point.
	gsmSCF → gprsSSF
Activity test GPRS	Used to check if the gsmSCF and the gprsSSF relationship is in existence. Appropriate actions are taken to recover the relationship if no response is received.
Applying charging GPRS	Used to instruct the gprsSSF to apply charging mechanisms to control the charging of a GPRS session or a PDP context.
Applying charging report GPRS ack	Response to the IF applying charging GPRS.
Cancel GPRS	Request to the gprsSSF to cancel all EDPs and report.
Connect GPRS	Used to request the gprsSSF to modify the APN that established a PDP context.
Continue GPRS	Request to the gprsSSF to continue the processing that had previously been suspended.
Entity released GPRS ack	Response to the IF entity released GPRS.
Event report GPRS ack	Response to the IF event report GPRS.
Furnish charging information GPRS	Request to include information in the CAMEL-specific logical call record.
Release GPRS	Instruction by the gsmSCF to tear down an existing GPRS session or PDP context.
Request report GPRS event	Instruction to monitor an event and send a notification on event detection.
Reset timer GPRS	Used to reset gprsSSF timer.
Send charging information GPRS	Used to send an e-parameter to the gsmSSF. This includes advice of charge.
	HLR → SGSN
Delete subscriber data	Instruction to delete CAMEL subscription data from an SGSN.
Insert subscriber data	Used by the HLR to insert subscriber data in the SGSN.
	SGSN → HLR
Insert subscriber data ack	A response to IF insert subscriber data.
Update GPRS location	Used by the SGSN to indicate to the HLR the CAMEL phases supported by the SGSN.

TABLE 8-14   CAMEL Information Flow for Short Message Service

CAMEL operation	Description
gsmSSF/gprsSSF → gsmSCF	
Event report SMS	Notification of an event to the gsmSCF. A previous IF, request report SMS event, triggers this event monitoring.
Initial DP SMS	Request for the instructions from the gsmSCF on detection of a trigger at DP.
gsmSCF → gsmSSF/gprsSSF	
Connect SMS	Used to request gsmSSF/gprsSSF to perform the actions to route the short message to a specific destination (for MO-SMS) or to deliver the short message to the MS.
Furnish charging information SMS	Request to include information in the CAMEL-specific logical MO-SMS or MT-SMS record.
Release SMS	Request to tear down an existing SMS transfer.
Request report SMS event	Instruction to start monitoring an event and send notification when the event is detected.
Reset timer SMS	Instruction to reset a gsmSSF/gprsSSF timer.
HLR → VLR/SGSN	
Delete subscriber data	Used by the HLR to delete CAMEL subscription data from a VLR/SGSN.
Insert subscriber data	Used by the HLR to insert CAMEL subscription data in a VLR/SGSN.
VLR/SGSN → HLR	
Insert subscriber data ack	A response to the IF insert subscriber data.
Update GPRS location	Used by the SGSN to introduce to the HLR the CAMEL phases and CAMEL Phase 4 CSIs offered by the SGSN.
VLR → MSC	
Continue CAMEL SMS handling	Instruction to the MSC to continue CAMEL-specific handling.
Send info for MO-SMS ack	Used to transport MO-SMS related subscription data from the VLR to the MSC.
MSC → VLR	
Send info for MT-SMS	Used to request the VLR to provide information to handle an MT-SMS.

CAMEL services. It contains the trigger information that is required to invoke CAMEL service logic for a roamer-originated call.

3. If the O-CSI is present, the VMSC suspends the call and examines O-CSI to get the gsmSCF address to establish a CAMEL control relationship.

TABLE 8-15   CAMEL Information Flow for SS Notification

CAMEL operation	Description
	MSC → gsmSCF
SS invocation notification	MSC uses this to notify the gsmSCF of a supplementary service invocation.
	HLR → VLR
Delete subscriber data	The HLR uses this to delete subscription data from a VLR.
Insert subscriber data	The HLR uses this to update a VLR with certain subscriber data.
SS invocation notification	The HLR uses this to notify the gsmSCF of a supplementary service invocation.
	VLR → MSC
Invoke SS result	The VLR uses this to transfer SS-CSI to the MSC. This is sent on successful invocation of supplementary services ECT and MPTY.
Send info for incoming call ack	The VLR uses this to send SS-CSI to the MSC. This is sent on successful invocation of supplementary service CD.

4. The VMSC (gsmSSF) sends an **initial DP** with an armed BCSM DP event as collected info to request instructions from the gsmSCF. This information flow also includes other mandatory information elements such as the service key and the called and calling party numbers.

5. The gsmSCF instructs the VMSC by sending the a **request report BCSM event'** (RRBE) to monitor for a call-related event and notify back.

6. The gsmSCF sends an **advice of charge** for that particular leg of a call, using **send charging info**.

7. The gsmSCF then sends a request to include specific call-related information in the final CDR generated for this call, using **furnish charging information**.

8. The gsmSCF sends instructions to the gsmSSF on the charging mechanism for this call using **apply charging**. This includes information such as the maximum call duration time after which the call shall be released and the tariff switch time until the next tariff switch applies.

9. The gsmSCF then requests the gsmSSF to continue the call processing and connect the call, using routing information included in **connect**.

10. The VMSC sends ISUP **IAM** to the destination PLMN (PLMN-C in this case).

TABLE 8-16   CAMEL Information Flow for Mobility Management

CAMEL operation	Description
VLR/SGSN → gsmSCF	
Mobility management event notification	Used to notify the gsmSCF of a mobility management event.
SGSN → HLR	
Update GPRS location	Used by the SGSN to provide information about supported CAMEL phases to the HLR.
VLR → HLR	
Update location	Used by the VLR to provide information about supported CAMEL phases to the HLR.
VLR → HLR	
Restore data	Used by the VLR to provide information about supported CAMEL phases to the HLR.
HLR → VLR/SGSN	
Delete subscriber data Insert subscriber data	The HLR uses this to delete subscription data from a VLR. The HLR uses this to update a VLR with certain subscriber data.

11. The VMSC send an **alerting** message to A on receiving ISUP **ACM** from the PLMN-C.

12. On receiving the ISUP **ANM** message from PLMN-C, the VMSC sends a **connect** message to the MS and establishes a connection to enable speech conversation.

13. The VMSC **contacts** the gsmSCF to report an O answer event and to get further instructions.

14. During the call, the gsmSSF may send an **apply charging report** to the gsmSCF. This may include information as to whether the call leg should be released because of maximum call duration time expiry.

15. If any of the parties involved releases the call, the VMSC coordinates with the destination PLMN and release all the resources. It reports back to the gsmSCF an O disconnect event. It then sends a final charge report to the gsmSCF, using **apply charging report**.

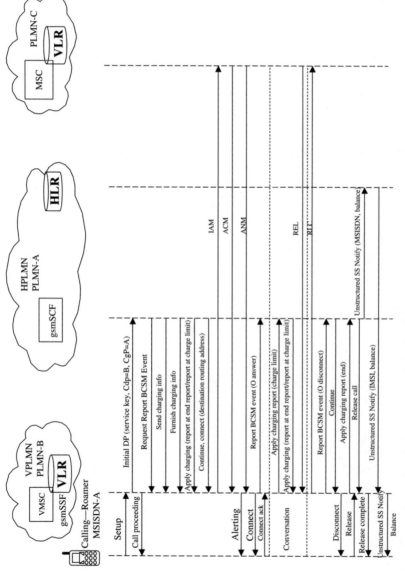

Figure 8-15 Procedure for a roamer-originated call.

16. The gsmSCF invokes a USSD procedure with the HLR to send credit balance information to the prepaid roamer.

Bibliography

3GPP TS 29.002, Mobile Application Part (MAP) specification.
3GPP TS 22.090, Unstructured Supplementary Service Data (USSD)—Stage 1.
3GPP TS 23.090, Unstructured Supplementary Service Data (USSD)—Stage 2.
3GPP TS 24.090, Unstructured Supplementary Service Data (USSD)—Stage 3.
3GPP TS 23.078, Customised Application for Mobile Network Enhanced Logic (CAMEL) Phase 4, Stage 2, Release 6.
3GPP TS 29.078, CAMEL Application Part (CAP) specification.

9

Inter-PLMN Roaming Testing

Enabling and managing inter-PLMN roaming is quite a challenging task. The service availability and QoS of roaming services depends not only on the home network but also on the visited network and all the intermediate international carrier networks. The building blocks of the HPLMN and the partner networks may be based on technology offered by different vendors and may be supporting different versions of technology standards. The routing tables in all participating networks need to be updated continuously to accommodate the changes in the network as a result of subscriber growth and the introduction of new services. A comprehensive inter-PLMN roaming testing is required before launching services to ensure service accessibility and availability across the network. Once the roaming service is launched, it is necessary to continuously monitor the QoS to identify and isolate faults proactively.

9.1 Overview of IREG Testing

The International Roaming Expert Group (IREG), as part of the GSM Association, has released a series of test specifications to ensure correct operation of roaming between two participating GSM network operators. GSM Association Permanent Reference Documents (PRDs) define the tests required. All the participating operators are bound by the terms of the GSM MoU.

The test structure is based on four steps. These steps are performed before roaming service launch.

Stage 1. MAP interface self-certification testing.

Stage 2. Internetwork SCCP and IP connectivity testing.

Stage 3.   Exchange of numbering and addressing data. Fault reporting, operation, and maintenance procedures.

Stage 4.   End-to-end functional capability testing.

In addition, billing verification is required for both inbound and outbound roaming service usage.

The purpose of breaking down the testing into stages is that Stage 1 tests need be done only once for each type and manufacturer of network element. These are very complex and thorough tests. The Stage 2 and Stage 3 tests need to be performed only when there are significant changes in the network. The Stage 4 tests must be repeated with each PLMN occasionally. It must therefore be a high-level functional test stage, and has to be completed in a short period.

Once the service is launched, then end-to-end functional testing (Stage 4) may be performed on a regular basis to ensure that ongoing network changes and newly introduced services are not resulting in any service disruption or degradation. These tests may also need to be conducted on request from partner networks as and when need arises.

9.1.1   Internetwork connectivity

In order to enable roaming between two partner networks, the following connectivity is established.

For GSM roaming:

- CCS7 SCCP/MAP connectivity to link the HPLMN HLR and the VPLMN VLR.

- International circuit-switched connection for transport of speech and circuit-switched data between partner networks.

Point-to-point connections are not feasible because of the cost involved in setting up and maintaining links. Usually, the wireless service provider utilizes the service offered by established international carriers for cost-efficient internetwork connectivity. The services offered by international carriers may vary in capabilities. In general, these include:

- Signaling access

- SCCP routing

- Online reporting

- Signaling conversion for interstandard roaming, if required

- SMS interworking

- Roaming management

For GPRS/3G roaming:

- CCS7 SCCP/MAP connectivity to link the HPLMN HLR and VPLMN VLR/SGSN.
- International circuit-switched connection for transport of speech or circuit-switched data between partner networks.
- International packet-switched connection between partner networks.

For packet-switched connection, wireless service providers utilize the services offered by global roaming exchange (GRX) operators. The services offered by GRX operators may vary in terms of capabilities. In general, these include:

- Routing of data packets and DNS queries between operators and their roaming partners (supporting GTP tunneling on both UDP and TCP as required by GTP specifications).
- DNS root services.
- Dynamic exchange of routing information between connected GPRS networks using BGP-4 routing protocol capabilities.
- Data security against spoofing and intrusion from the public network.
- Service level guarantees.

9.1.2 End-to-end functional testing

PRD IR.24 defines an end-to-end functional capability specification for inter-PLMN roaming (Stage 4 testing). Stage 4 testing means that the tests are restricted to the top-level capabilities only, i.e., no negative or provocative test cases included. The correct operation of roaming is verified by simulating the inbound roamers. The participating networks exchange test SIMs to carry out these tests. PRD IR.24 defines the tests for MAP interworking and basic circuit-switched (CS) services. These tests are applicable for 2G and 3G CS environments. The tests for GPRS (2.5G) and 3G packet-switched (PS) services are defined in PRD IR.35. PRD IR.26, an addendum to IR.24, defines tests for supplementary services and operator-determined barring. PRD IR.27 defines the tests for circuit-switched data services, Group 3 fax, and alternate fax/speech. PRD IR.32 defines the tests, which are performed to ensure CAMEL interworking for prepaid roaming.

IR.24, IR.26, and IR.27 tests are resource-intensive if the operators need to perform these tests manually. In the present context, where the number of roaming partners could be significantly large, it may not be practical to conduct tests on a regular basis for each partner network. PRD IR.28 describes the tools and mechanism to automate the basic tests.

PRD IR.29 describes the minimum functionality for the test equipment required to automate the tests.

PRD TD.06 describes the billing and accounting data, which is exchanged in conjunction with these tests if required for the testing.

9.2   IREG Testing—Test Setup

9.2.1   GSM roaming test setup

Figure 9-1 shows the basic test setup for IREG testing. PLMN-B is shown as the network from where the tests are invoked for inbound roamers, i.e., subscribers from PLMN-A visiting PLMN-B. However, the same tests need to be repeated by PLMN-A for its inbound roamers, i.e., subscribers from PLMN-B.

Test mobile station MS1-A is a PLMN-A subscriber roaming in PLMN-B and is equipped with the SIM1-A. The second test mobile station MS2-A is also a PLMN-A subscriber and is equipped with SIM2-A.

TP1-A, TP2-A, TP1-B, and TP2-B are fixed line test phones. TP1-A and TP2-A are the subscribers of a fixed line operator in the same country as PLMN-A. TP1-B and TP2-B are the subscribers of a fixed line operator in the same country as PLMN-B.

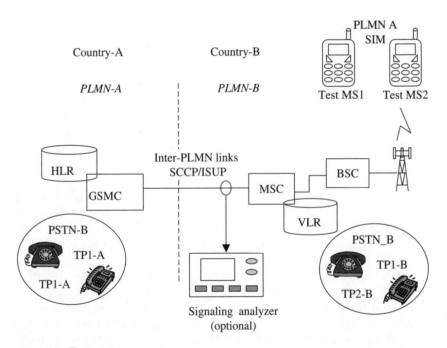

Figure 9-1   Test setup.

A minimum of two SIMs are supplied by the PLMN-A to the PLMN-B and vice versa. The following information is needed to enable the tester to configure two test mobiles for testing:

- Personal identification number (PIN)
- Personal user key/pin unblocking key (PUK)
- Super PIN
- MSISDN
- IMSI
- Basic services subscription information
- Supplementary services subscription information

A signaling tester with CCS7 support is optional test equipment. This is required to log the call and the transaction traces, which are used for localizing and diagnosing the faults in case of a test failure.

The participating networks exchange the PRD IR.21 document, which lists all the necessary network information required to enable roaming. It is mandatory to update and exchange this document, whenever necessary, to reflect changes in the network. The receiving network updates routing tables of all relevant network elements such as GWMSC according to the information supplied with PRD IR.21. The GSM-specific information includes:

- E.164 number series, e.g., MSISDN range, MSRN range, GT list for MSCs, and SMSCs
- E.212 number series, i.e., MCC and MNC
- E.214 mobile global title (MGT), i.e., CC and NC
- IMSI to MGT translation rules
- Number portability support, if available
- International SCCP gateway point codes
- Application context supported version

9.2.2 GPRS roaming test setup

A similar setup as described in Section 9.2.1 is required for end-to-end functional testing for inter-PLMN GPRS roaming.

Figure 9-2 shows a typical test setup for the GPRS roaming testing. The test mobile phones, i.e., MS1 and MS2, are GPRS class A/B/C phones. The following information is needed to enable tester to configure test mobiles for testing:

- Personal identification number (PIN)
- Personnel user key/pin unblocking key (PUK)

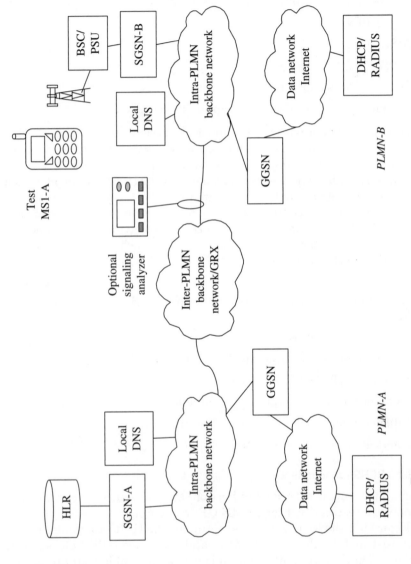

Figure 9-2 Test setup.

TABLE 9-1   Test SIM Profiles

SIM	APN	VPLMN allowed flag
1	HPLMN's APN	Yes
2	HPLMN's APN	No
3	Wild card "*"	Yes

- Super PIN
- MSISDN
- IMSI
- Basic services subscription information
- Subscriber data related to GPRS attach and PDP contexts
- Access point names

To test different PDP context scenarios, the supplied test SIMs are of three different profiles set in the HLR (Table 9-1).

A signaling tester to log the signaling transactions, service activation, and user data is desirable. This is required to verify correctness of signaling and data exchange between PLMNs. This test equipment is also very valuable in case of failure to isolate and diagnose the faults. The signaling tester shall support GSM/GPRS/3G interfaces and protocols.

PRD IR.21 has been extended to include necessary information to enable inter-PLMN GPRS roaming. The additional information required is:

- Supported BSSAP+ application contexts and version
- APN operator identifier
- Primary and secondary DNS IP address
- Inter-PLMN GSN backbone IP address ranges
- PDP context types and profiles
- Security-related information, if any

9.3   IREG Tests

9.3.1   GSM basic service tests

The objective of these tests is to ensure that a roamer from a roaming partner network can successfully register in the network under test. It also verifies whether the HLR in the home network updates the serving VLR with subscriber subscription data. Once the registration is complete, the cancel location service is invoked to verify that it operates correctly.

Location Update. The purpose of this test is to confirm the correct operation of the MAP update location and insert subscriber data procedures.

Pretest conditions	■ The MS1-A is equipped with the SIM from PLMN-A. The SIM subscription data is set to include basic and supplementary services. ■ Delete MS1-A data from PLMN-B VLR, if any exists. ■ Set network selection in manual mode on MS1-A
Test actions	■ Power ON mobile. ■ Select network PLMN B.
Verification	■ MS1-A indicates a successful selection of PLMN-B. ■ A new record for MS1-A with the identical basic and supplementary services shall be created in PLMN-B VLR.

Optionally, the update location and insert subscriber data transactions can be traced by using a protocol analyzer. Figure 9-3 shows the signaling sequence flow for the update location procedure. This is useful to verify the supported MAP version, mandatory and optional information elements, and total response time. The transaction trace is also used to diagnose the problems if the UL procedure fails. The most commonly encountered faults are wrong routing and address translation.

Operator control of service

Cancel location on withdrawal of subscription.
The purpose of this test is to confirm the correct operation of the MAP cancel location procedure

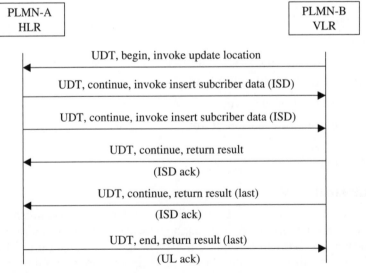

Figure 9-3 UL protocol sequence.

when the home network withdraws the services to its subscriber currently roaming in a foreign network. The HLR invokes the cancel location procedure. This results in deletion of the subscriber data in the VLR.

Pretest conditions	There exists a record of MS1-A in the PLMN-B VLR as a result of a previous successful update location operation.
Actions	Request PLMN-A tester to delete MS1-A subscription in the HLR, using appropriate MMI command.
Verification	Check PLMN-B VLR for MS1-A record. The record shall be erased from the VLR as a result of the cancel location operation.

Optionally, the cancel location procedure can be traced by using a protocol analyzer to verify all the parameters and to diagnose the problem in case of test failure. Figure 9-4 shows the signaling sequence flow for the cancel location procedure.

Operator-determined barring (ODB)—all calls. Operator-determined barring (ODB) allows the network operators to bar outgoing or incoming calls and also roaming services. This feature provides greater control to network operators. For example, a network operator, to avoid losses, could bar a suspected fraudulent roamer. The purpose of this test is to confirm that the operator-determined barring applies correctly when a subscriber is roaming.

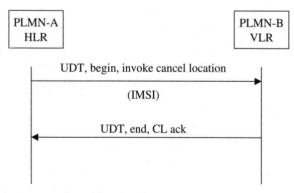

Figure 9-4 Cancel location.

Pretest conditions	There exists a record of MS1-A in the PLMN-B VLR as a result of a previous successful update location operation. The ODB data confirms that the roamer is allowed to make outgoing calls and receive incoming calls while roaming.
Test actions	Request PLMN-A tester to activate ODB to bar all outgoing and incoming call using an appropriate MMI command.
Verification	■ Verify that the PLMN-B VLR data contains modified ODB information for MS1-A.
	■ An attempt to establish an outgoing call using MS1-A shall fail.
	■ An attempt to call MS1-A shall fail.
	■ MS1-A shall be able to send and receive SMS.

Roamer-to-roamer call in a visited network. The purpose of this test is to confirm that a roamer can successfully initiate a call. The called party used for this test is another subscriber from PLMN-A roaming in the same network. It also tests roamer-terminated calls and verifies the provide roaming number procedure (PRN).

Pretest conditions	Both roamers MS1-A and MS2-A have completed the update location procedure successfully.
Test actions	MS1-A initiates a call with the called party as MS2-A. Hold the call for 1 minute or more.
Verification	■ Verify that the call successfully gets established.
	■ Assess the voice quality of service and use of echo cancellation.
	■ Optionally, provide roaming number (PRN) procedure and ISUP call legs can be viewed and verified using a protocol analyzer. The transaction and call trace is used to further diagnose the reason in case of call failure.

Roamer-terminated calls

The VLR has no record of the roamer. The purpose of this test is to ensure that calls to a roamer that has previously registered in PLMN-B, and for which no data exist in the VLR, is either successfully terminated or an appropriate announcement is fed.

Pretest conditions	The HLR shows that MS1-A has roamed in PLMN-B. MS1-A has previously registered in PLMN-B but there is no data in the VLR related to MS1-A (e.g., IMSI purge because of inactivity for an extended period of time). The VLR has no location updated since the MS1-A record in the VLR was removed.
Test actions	PSTN phone TP-B calls MS1-A.
Verification	There are two possible scenarios in this case:
	■ The call is successful. In this case, the VLR is able to recover by using the send parameter procedure with the HLR. Check the speech quality and echo.
	■ The call fails within 30 seconds with an appropriate announcement.
	In both cases the result is successful.

Roamer has performed IMSI detach. The purpose of this test is to ensure that calls to a roamer that has been detached or deregistered in PLMN-B, and for which a record exists in the VLR, are fed with an appropriate announcement.

In response to a provide roaming number request from the HPLMN for the visitors who have performed IMSI detach, the VPLMN returns an error code indicating absent subscriber. Figure 9-5 shows the call sequence for this test case.

Pretest conditions	Ensure that the MS1-A is detached in the VLR.
Test actions	TP-B attempts to call MS1-A.
Verification	The call fails to terminate; an appropriate announcement is fed by the PLMN-A GMSC to the calling subscriber, i.e., TP-B.

Roamer is not able to respond to paging. The purpose of this test is to verify that the VPLMN provides a correct announcement to a calling party when it attempts to call a roamer in a visited network that is unable to respond to a paging message.

Pretest conditions	Ensure that the MS1-A is disabled (but not detached).
Test actions	TP-B attempts to call MS1-A.
Verification	The call fails to terminate; an appropriate announcement is fed by PLMN-B to the calling subscriber, i.e., TP-B.

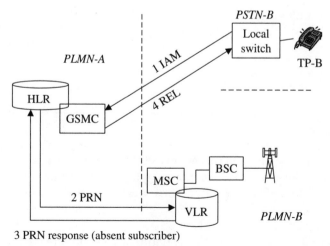

Figure 9-5 Call sequence for IMSI detach.

9.3.2   GSM supplementary services tests

The purpose of these tests is to verify that an MS is able to activate or deactivate supplementary services while roaming in a foreign network.

Barring of all outgoing calls.   This supplementary service allows a roamer to bar all outgoing calls. The roamer may use this service to control calls, costs, and possible misuse by others.

Pretest conditions	As a result of a roamer-initiated barring action, the HPLMN HLR entry contains SS: BAOC active. This means subscriber is barred for all outgoing calls except emergency services.
Test actions	The MS initiates one call to an emergency number. The MS initiates a second call to a local number.
Verification	Calls to the emergency number shall succeed. Outgoing calls to numbers other than the emergency number shall fail.

Barring of outgoing international calls.   This supplementary service allows a roamer to bar all outgoing international calls. The roamer may use this service to control calls, costs, and possible misuse by others.

   The purpose of this test is to verify that a roamer is not able to make an international outgoing call if this service has been activated.

Pretest conditions	As a result of a roamer-initiated barring action, the HPLMN HLR entry contains SS: BOIC active.
Test actions	MS1-A calls to TP-B or PSTN-B operator position. MS1-A attempts an outgoing call to the home PLMN. MS1-A attempts to call to an international number in a third country.
Verification	The call to the operator position shall be successful but the call to the HPLMN or any other international number shall fail.

Barring of outgoing international calls except to HPLMN country.   This supplementary service allows a roamer to bar outgoing international calls except those directed to the home country. The roamer may use this service to control calls and costs. This feature is network dependent and is not necessarily supported by a VPLMN.

   The purpose of this test is to verify that a roamer in a visited network can activate the supplementary service: barring outgoing calls except to HPLMN.

Pretest conditions	As a result of a roamer-initiated barring action, the HLR entry contains SS: BOIC ex HC active.
Test actions	1. MS1-A attempts a call to the PLMN-A number (home country).
	2. MA1-A attempts a call to the PSTN-A (home country)
	3. MS1-A attempts a call to the PLMN-B or PSTN-B number (visited country).
	4. MS1-A attempts a call to any other third country.
Verification	The call attempt to any country other than home or visited shall fail. If the VPLMN does not support this feature, then the serving MSC makes a functional fall back and acts as if BOIC (see "Barring of Outgoing International Calls except to HPLMN Country," above) is active. Call attempts 1, 2, and 3 listed under test action shall succeed while call attempt 4 shall fail.

Barring of all incoming calls when roaming. This supplementary service allows a roamer to bar all incoming calls when roaming outside the HPLMN country. The roamer may use this service to control their calls and costs. The purpose of this test is to confirm the support of the activation of this service in a VPLMN.

Pretest conditions	As a result of a roamer-initiated barring action, the HLR entry contains SS: BAIC roaming active.
Test actions	TP-B calls MS1-A.
Verification	The call shall fail.

Call forwarding on not reachable. This service allows the mobile subscriber to forward all incoming calls when the mobile subscriber is not reachable. Some networks implement certain restrictions to avoid possible fraudulent use of this feature. For example, the HLR may not allow registering a forwarded call to a number outside the HPLMN. The serving MSC may also prevent call forwarding to another country. The behavior of this service is tested in two scenarios.

1. Before IMSI detach

2. After IMSI detach

The battery of an MS is removed when it is ON and attached to a network to simulate a situation where a MS is not reachable and IMSI is not detached.

Pretest conditions	As a result of a roamer- (MS1-A-) initiated call forwarding on not reachable, the HLR entry for MS1-A contains SS: CFNRc active. The forwarded-to address is TP1-B. In cases where HLR does not allow call forwarding to another country, the forwarded-to address can be TP1-A or MS2-A. The MS2-A is another PLMN-A subscriber roaming in the PLMN-B.
Test actions	TP2-B attempts to call to MS1-A.
Verification	The call shall be forwarded to TP1-B (or TP1-A or MS2-A). Verify the speech quality and check for echo cancellation.

Call forwarding on busy.   This service allows the mobile subscriber to forward all incoming calls when the mobile subscriber is busy.

Pretest conditions	As a result of a roamer- (MS1-A-) initiated call forwarding on busy, the HLR entry for MS1-A contains SS: CFB active. The forwarded-to address is TP1-B. In cases where HLR does not allow call forwarding to another country, the forwarded-to address can be TP1-A or MS2-A. MS2-A is another PLMN-A subscriber roaming in the PLMN-B.
Test actions	■ MS1-A establishes a call with any other local or international number and remains in conversation. ■ TP2-B attempts to call to MS1-A.
Verification	The call shall be forwarded to TP1-B (or TP1-A or MS2-A). Answer the call and verify the speech quality and check for the echo cancellation.

Call forwarding on no reply.   This service allows the mobile subscriber to forward all incoming calls when the mobile subscriber does not reply.

Pretest conditions	As a result of a roamer- (MS1-A-) initiated call forwarding on no reply, the HLR entry for MS1-A contains SS: CFNRy active. The forwarded-to address is TP1-B. In cases where HLR does not allow call forwarding to another country, the forwarded-to address can be TP1-A or MS2-A. The MS2-A is another PLMN-A subscriber roaming in the PLMN-B.
Test actions	TP2-B attempts to call MS1-A. Do not answer MS1-A.
Verification	The call shall be forwarded to TP1-B after few rings (or TP1-A or MS2-A). Answer the call, verify the speech quality, and check for echo cancellation.

9.3.3   GSM SMS tests

Mobile-originated and -terminated SMS.   The purpose of this test is to verify that a roamer can transmit and receive an SMS while in a foreign network.

Pretest conditions	MS1-A and MS2-A are successfully registered in a visited network, i.e., PLMN-B.
Test actions	■ Switch OFF MS2-A. ■ Edit an SMS on MS1-A and send it to MS2-A. ■ Switch ON MS2-A.
Verification	Verify that MS2-A receives the SMS and the contents of the SMS are intact.

9.3.4 GSM SS and ODB tests

Call hold and call waiting

Call hold. The call hold service allows a mobile subscriber to interrupt communication on an existing active call and then subsequently reestablish communication.

The purpose of the test is to verify that a roamer can invoke call hold supplementary service in a visited network.

Pretest conditions	Supplementary service call hold is provisioned in the HLR for the roamer (MS1-A with SIM1-A) under test.
Test actions	▪ Switch ON MS1-A and select PLMN-B as a roaming network. ▪ Call MS1-A, using a fixed line PSTN-B. ▪ MS1-A puts call on hold. ▪ MS1-A retrieves call from held state. ▪ MS1-A resumes the call and is able to terminate call.
Verification	This verifies that MS1-A, roaming in PLMN-B, is able to invoke the call hold feature.

Call waiting. The call waiting service allows the mobile subscriber to be notified of an incoming call while engaged in an active or held call.

The purpose of this test is to verify that a roamer can invoke call waiting supplementary service in a visited network.

Pretest conditions	Supplementary service call waiting is provisioned in the HLR for the roamer (MS1-A with SIM1-A) under test.
Test actions	▪ Switch on MS1-A and select PLMN-B as a roaming network. ▪ Ensure that MS1-A is busy. ▪ Call MS1-A, using a fixed line PSTN-B. ▪ MS1-A gets an indication of call waiting. ▪ MS1-A clears the first call and connects the call from PSTN-B.
Verification	This verifies that MS1-A roaming in PLMN-B is able to invoke the call waiting feature.

Multiparty call. This supplementary service provides a mobile subscriber with the ability to have a simultaneous communication with more than one party. This is a useful service for mobile users and generates additional revenues for the operators.

The purpose of this test is to verify that the MS roaming in a visited network is able to invoke a multiparty call.

Pretest conditions	Supplementary service multiparty is provisioned in the HLR for the roamer (MS1-A with SIM1-A) under test.
Test actions	■ MS1-A roaming in a visited network calls a fixed line phone. ■ MS1-A holds the call. ■ MS1-A calls another fixed line. ■ MS1 adds the held party to the connected party.
Verification	This verifies that MS1-A, roaming in PLMN-B, is able to invoke the multiparty call feature.

Advice of charge. This supplementary service permits the mobile station to display an estimate of the bill that will be levied in the HPLMN.

Pretest conditions	The HLR entry for MS1-A contains SS: AoCC provisioned. MS1-A is registered in the visited PLMN-B.
Test actions	■ Initiate a mobile-originated call ■ Initiate a mobile-terminated call ■ Make a note of charge advice information (CAI) before and after the call.
Verification	Verifies that CAI for mobile-originated and -terminated call indicates the true cost of the calls made.

Calling line identity presentation/restriction. Calling line identification presentation (CLIP) provides the called party with the possibility of receiving the calling line's identity (MSISDN number of calling party). Calling line identification restriction (CLIR) enables the calling party to prevent presentation of its line identity to the called party.

The availability of these features depends on the provisioning of the HPLMN, the VPLMN, and the international network.

The purpose of this test is to ensure that calling line identity is presented or restricted according to the provisioning.

Pretest conditions	The HPLMN and the VPLMN supports CLI service. The HLR entry for MS1-A contains SS: CLIP provisioned and SS: CLIR permanent mode: provisioned. The HLR entry for MS2-A contains SS: CLIP provisioned and SS: CLIR not provisioned.
Test actions	■ MS2-A calls MS1-A. ■ MS1-A calls MS2-A.
Verification	In the first case, MS1-A displays calling line identity, i.e., MS2-A MSISDN number. In the second case, MS2-A does not display calling line identity.

Operator-determined barring. This provision allows an HPLMN service provider to control its roaming subscriber access to all or certain categories of services. The operators use this capability to limit the financial risk and stop misuse of services by suspected, fraudulent, or nonpaying subscribers. The operator can initiate ODB any time. The ongoing call gets terminated and future access to the service is barred. The HPLMN HLR invokes the insert subscriber data procedure to update the serving VLR on changes.

The purpose of the test is to verify that the VPLMN takes action on an HPLMN request and bars the roamer from certain or all services. The table below applies to operator-determined barring of all outgoing calls. The same type of test is repeated for other service categories.

Pretest conditions	As a result of action taken by an operator to bar MS1-A from making outgoing calls, the HLR entry contains ODB: BAOC active.
Test actions	■ MS1-A attempts to call TP1-B or TP-1A or MS2-A. ■ MS1-A attempts to call an emergency number.
Verification	Calls to an emergency number succeed; calls to other numbers fail.

USSD.   The purpose of this test is to verify that a roamer in a visited network is able to access USSD services as provided by the HPLMN.

Pretest conditions	The HPLMN has implemented at least one USSD application and the tester in the VPLMN has information on the access code for the service and its behavior.
Test actions	MS1-A invokes the USSD application by typing the USSD string and sending it, e.g., *111*2345678# Send.
Verification	The test is successful if the HPLMN returns with the right response expected from the USSD application.

9.3.5   GPRS attach

The purpose of this test is to verify that a roamer can perform GPRS attach in a visited network under test. Successful GPRS attach also concludes that the essential MAP procedures such as update GPRS location and insert subscriber data are successful.

Pretest conditions	No valid MM contexts exist in the HLR for the MS1-A.
Test actions	Switch ON MS1-A in PLMN-B and perform GPRS attach.
Verification	In the response GPRS attach accepted is received.

9.3.6   GPRS PDP context tests

A roamer, once successfully attached to a network, can initiate a PDP context to access services offered by the VPLMN or the HPLMN. The possible scenarios are as follows.

1. Intranet access using the home GGSN

2. Internet/ISP access using the home GGSN

3. Internet/ISP access using the visited GGSN

Test case 1: Intranet access using home GGSN. The purpose of this test is to verify that a roamer is able to successfully access an intranet in a visited network, using the home GGSN (HGGSN). In this case, the user is not allowed to access data services using visited GGSN (VGGSN).

Figure 9-6 shows the steps to successfully activate PDP context. The MS sends the activate PDP context request with a valid and subscribed APN (say, 'Intranet') to the serving SGSN in the visited network. The VSGSN appends the PLMN-A APN operator ID and queries the DNS server to get the GGSN address. The VSGSN then sends a create PDP context message to the HGGSN. On successfully obtaining the create PDP context from the HGGSN, the VSGSN returns an activate PDP context accepted message with a valid PDP type and PDP address to the MS.

Pretest conditions	MS1-A has subscribed to the APN "Intranet." The VPLMN not allowed flag is set for the APN. MS1-A is GPRS attached in PLMN-B.
Test actions	MS1-A performs an activate PDP context request. A request to download a specified file is made, using a supported transport protocol (HTTP, FTP, or WAP).
Verification	The file transfer completes successfully with no errors.

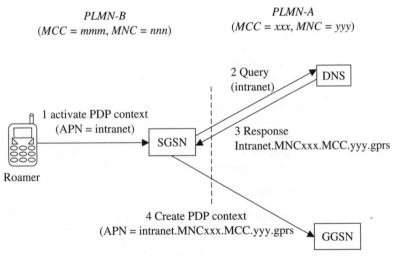

Figure 9-6   Intranet access using HGGSN.

Test case 2: Internet/ISP access using home GGSN.   This test case is similar to test case 1, except that MS1-A uses a subscribed APN to access the ISP.

Test case 3: Internet/ISP access using visited GGSN.   The purpose of this test is to verify that a roamer is able to successfully access the ISP in a visited network, using the VGGSN. In this case, the user is allowed to use the VGGSN.

Figure 9-7 shows the steps to successfully activate PDP context. The MS sends an activate PDP context request with a valid and subscribed APN (say, "Internet") to the serving SGSN in the visited network. The VSGSN appends PLMN-B operator ID and queries the local DNS server to get the GGSN address. The VSGSN then sends a create PDP context message to VGGSN. On receiving a successful create PDP context response from the VGGSN, the VSGSN returns an activate PDP context accepted message with a valid PDP type and PDP address to the MS.

Pretest conditions	MS1-A has subscribed to the APN "Internet." The VPLMN not allowed flag is not set for the APN. MS1-A is GPRS attached in PLMN-B.
Test actions	MS1-A performs an activate PDP context request with the APN "Internet." A request to download a specified file is made using a supported transport protocol (HTTP, FTP, or WAP).
Verification	The file transfer completes successfully with no errors.

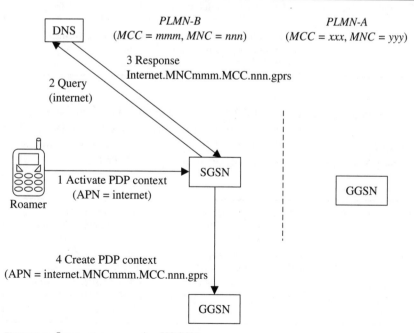

Figure 9-7   Internet access using VGGSN.

9.3.7 GPRS SMS tests

The purpose of this test is to verify that a roamer in a visited network is able to send and receive an SMS, using GPRS bearer.

Pretest conditions	MO-/MT-SMS is provisioned in the HLR subscription, including GPRS subscription data for both MS1-A and MS2-A. Both MS1-A and MS2-A are OFF.
Test actions	■ Switch ON MS1-A and perform GPRS attach only. ■ Create an SMS and send it to MS2-A. ■ Switch ON MS2-A and perform GPRS attach only.
Verification	MS2-A receives SMS successfully with no error.

9.3.8 GPRS operator control of service

Location cancellation for a roamer.   This is used by the HPLMN to delete subscription information from the HLR. If the subscriber is currently roaming in a visited country, then the subscriber record in the serving SGSN and the GGSN is erased and PDP contexts are deleted.

Pretest conditions	MS1-A is GPRS attached with SGSN-B, and PDP context is activated.
Test actions	The HPLMN tester deletes the MS1-A subscription from the PLMN-A HLR, using a man-machine interface.
Verification	Verify that the SGSN-B and GGSN-A/GGSN-B records for MS1-A have been erased and PDP context has been deactivated.

Operator-determined barring—PDP context activation.   This provision allows an HPLMN service provider to bar a roamer access to GPRS services. The operators use this capability to limit the financial risk and stop misuse of services by suspected, fraudulent, or nonpaying subscribers.

Pretest conditions	MS1-A is GPRS attached only to PLMN-B. The PLMN-A tester activates ODB for MS-initiated PDP context activation for MS1-A.
Test actions	■ MS1-A attempts to transmit an SMS to MS2-A. ■ MS1-A attempts PDP context activation.
Verification	MS2-A receives SMS with no error. The SGSN-B record for MS1-A contains ODB information and no PDP context activation.

9.3.9 Other tests

In addition to services mentioned here, GSM circuit-switched data and fax services are also supported by many operators for roamers. A special setup is required to perform a data services test. It consists of a wireless modem with a PC running the test program. The PC transmits test strings to the receiving PSTN modem. In the case of data call, the ASCII character string 01 is transferred 1000 times. In the case of a fax call, ITU-T telefax test charts 1 and 2 are transferred.

The test scenarios for data calls are as follows;

1. Mobile-originated data call from MS1-A to PSTN-A
2. Mobile-terminated data call from PSTN-A to MS1-A

The tests are performed for data transfer rates of 2400, 4800, and 9600 bits/s in both transparent and nontransparent modes.

The test scenarios for FAX calls are as follows;

1. Mobile Originated FAX call from MS1-A to PSTN-A
2. Mobile Terminated FAX call from PSTN-A to MS1-A

The tests are performed for a data transfer rate of 9600 bits/s in transparent mode.

9.4 IREG Tester

At this time, there are more than 400 wireless service providers operating GSM networks. Most of them are also migrating to enhanced data services using GPRS or 3G technologies. International roaming is one of the most popular services and is offered by almost all the operators. In order to have global reach, they need to sign up with as many partners as possible. As the numbers of IREG tests are significant, it is not feasible to perform IREG testing for each partner manually on a regular basis. Thanks to commercially available IREG testers, most of the IREG tests can be automated.

IREG testers use test phones or modems as test interfaces to simulate users. The test SIMs from the partner networks are inserted in test phones or modems. Depending on vendor specification, IREG testers interface with networks on the Um, A, or Gb interface. Usually, IREG testers support multiple test interfaces. Figure 9-8 shows IREG testers connected to Um, A, and Gb interfaces.

IREG testers typically consist of the following functional modules:

■ *Resource management:* Manages the common resources in a multiuser environment where several users may have test scripts running.

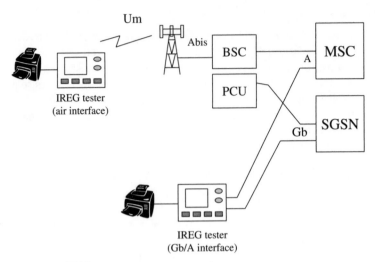

Figure 9-8 IREG testers.

- *Test initialization:* Initializes the test interface module to satisfy pretest conditions for a specific IREG test.
- *Test scheduler:* Schedules the test in a sequence defined by user.
- *Test execution:* Executes the predefined test and sends results to a database, printer, or user interface. The test may be performed on demand or periodically.
- *Database:* Stores the configuration and measurement results.
- *Reporting module:* Issues standard reports based on GSM MoU PRD IR documents. Additional reports for service availability, accessibility and performance may also be available in case periodic tests are performed.
- *User interface*

Typically, IREG testers support all the measurements defined in IR.24, IR.26, IR.27, IR.32, and IR.35 with the exception of those measurements that require human intervention.

Bibliography

GSM MoU Association PRD IR.24, End to End Functional Capability—Specification for Inter-PLMN Roaming.

GSM MoU Association PRD IR.26, End to End Functional Capability—Specification for Inter-PLMN Roaming. Addendum for Phase 2 Supplementary—Services and Operator Determined Barring.

GSM MoU Association PRD IR.27, Functional Test Specification for Inter PLMN Roaming. Phase 1 Data Services, FAX Services.

GSM MoU Association PRD IR.28, Specification of the infrastructure in a PLMN to allow automatic testing.

GSM MoU Association PRD IR.29, Proposal of a minimal requirement on an automatic test equipment for roaming.

10

Roaming Service Management and Troubleshooting Faults

Roaming brings significant profit to a wireless service provider's bottom line. It is also one of the most popular services offered by service providers. Managing roaming services, however, is a challenge. There is minimal data available on roaming service from the network elements involved. Monitoring the network elements and associated signaling and data links for their availability status and performance provides a good indication of the network health but reveals very little about underlying services. In real-life situations, it is often observed that a roamer is not able to access a particular service although the network status shows no faults. For example, a network management center based on network element monitoring may not be able to generate an alert in case of an error in the routing table. An error in the routing table may result in roaming failure, as critical signaling messages such as update location may not be able reach HPLMN HLR. It is therefore necessary to evolve a different strategy for monitoring of services as complex as roaming.

Many times the disruption or degradation of roaming services goes unnoticed for a long time, resulting in loss of revenue, as roamers tend not to call customer care promptly. Inbound roamers, on encountering a problem, would rather move to another available network, as they have choice and no loyalty to any visited network, causing instant churn. Outbound roamers may not report back, as they are in a foreign country and report only after arrival to their home network. Therefore, it is important to monitor the roaming services proactively to detect degradation or disruption before the roamers notice it.

This chapter examines various key quality indicators that characterize roaming services and how to monitor then proactively to ensure minimum disruption.

10.1 Quality of Service—General Concepts

A typical user is not really aware and not concerned with how a particular service is designed and implemented. Generally, users express their degree of satisfaction in using a service in a nontechnical way. The user perception of quality is based simply on his/her experience in using services end to end. The network performance is a critical factor, but quality of service is not just about that. Other external factors, such as the mobile phone, service tariff, and customer care, all have a serious influence on user perception and resulting satisfaction. The question is how to measure the user perception on quality of service.

One of the most frequently used definitions for quality of service (QoS) is stated in ITU-T Recommendation E.800. It defines QoS from the end-user perspective and its relationship with the network performance.

> The quality of service is the collective effect of service performance, which determines the degree of satisfaction of a user of the service.

ITU-T E.800 further defines network performance as:

> The network performance is defined as the ability of a network portion to provide the functions related to communication between users.

Network providers measure the performance of their network portions or individual service components against the agreed key performance indicators (KPIs). By definition, the KPIs are network focused. The KPIs are very useful and important measurements for network operations, as they indicate the health of a service component. However, they alone cannot be used to specify a user's QoS requirements nor represent end-to-end service delivery quality. As the user is concerned only with the end product offered by service/network providers, new key quality indicators (KQIs) are defined to measure a specific aspect of the performance of the product and product components, i.e., services and service elements. The KQIs are derived from the KPIs and other data sources that may influence customer satisfaction.

Table 10-1 illustrates the distinction between network performance and quality of services as described in ETSI ETR 300 specification.

10.1.1 Service-independent quality indicators

Network accessibility. When a roamer tries to access the visiting network, the visited network authenticates the roamer with the home network and obtains roamer information by means of the update location procedure. The success of this action depends on the performance of both home and visited networks.

TABLE 10-1   NP and QoS

Network performance	Quality of service
Provider oriented, network focused	User oriented, service focused
Connection element attribute, e.g., throughput	Service attribute, e.g., speed
Focus on planning, design and development, operation and maintenance, e.g., error rate	Focus on user-observable effects, e.g., accuracy
End-to-end or network connection element capabilities, e.g., delay, delay variance	Between (at) service access points, e.g., dependability

- The visited network provides radio access and routing of required messages back to the home country.

- The home network updates the HLR with the new location and provides the necessary subscriber information to the visited network.

The UL success rate provides a good indication of the network coverage and security and is used as a measure of network accessibility for both GSM and GPRS networks from a roamer perspective. In GPRS, for example, it is indicated by ability to attach in the visited network.

10.1.2 Service-dependent quality indicators

Service accessibility. The service access quality is determined by the success rate of the services a roamer tries to access, such as establishing a voice call, receiving a voice call, and sending/receiving an SMS/MMS. The QoS as experienced by a user while using a service is termed *service integrity*. For example, speech quality in a voice call is an indicator of service integrity.

Service retainability. Service retainability describes the termination of services (in accordance with or against the will of the user). For example, once a user establishes a call and is in the conversation phase, it should remain in this phase till one of the parties decides to disconnect. However, in a real-life scenario, there is a possibility that the call will get terminated prematurely without the will of the users, e.g., because of blind spots in radio coverage.

10.1.3 Quality indicators by services

GSM services. Table 10-2 lists the key quality indicators (KQIs) for some popular GSM services used by a roamer. The formula to compute each KQI is also defined.

TABLE 10-2   GSM Services and KQIs

Services	Services access—KQIs	Abstract formula
Voice calls	Service accessibility	Service accessibility (%) = number of successful call attempts × 100/number of call attempts.
	Call setup time	See Figure 10-1.
	Speech quality	Speech quality is measured according to ITU-T Recommendation P.862. The quality is indicated in terms of MOS score.
	Call completion rate (CCR)	CCR (%) = number of intentionally terminated calls × 100/number of successful call attempts.
SMS	Service accessibility— MO-SMS	Service accessibility MO-SMS = number of successful SMS service attempts × 100/number of all SMS service attempts.
	Access delay— MO-SMS	See Figure 10-2.
	End-to-end delivery time	See Figure 10-3.
	Completion rate, SMS circuit switched	CS = (successful received test SMS— duplicate Test SMS—corrupted test SMS) × 100/number of all test SMS sent.
Circuit-switched data services	Service accessibility	SA-CS (%) = Number of successful call attempts × 100/number of call attempts.
	Setup time	See Figure 10-4.
	Data quality	The end-to-end data quality (DQ) is measured in terms of average data throughput in both uplink and downlink directions on best-effort basis. For conversational class data: DQ (uplink) = X bits/second DQ (downlink) = X bits/second For streaming class data: DQ (downlink) = X bits/second
	Completion rate	CR = number of calls terminated by end users/number of successful data call attempts.

GPRS services. Table 10-3 lists the key quality indicators (KQIs) for a few popular GPRS services used by a roamer. The formula to compute each KQI is also defined.

QoS parameter for bearer-independent data services. The end-to-end data quality for bearer-independent data services is measured in terms of

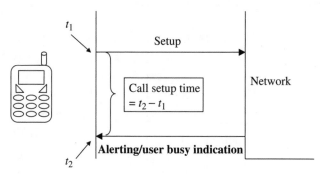

Figure 10-1   Call setup time for voice calls.

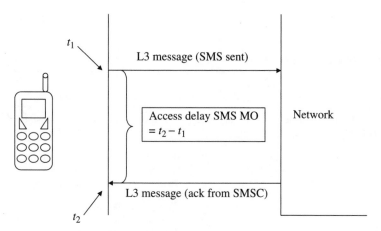

Figure 10-2   Access delay MO-SMS.

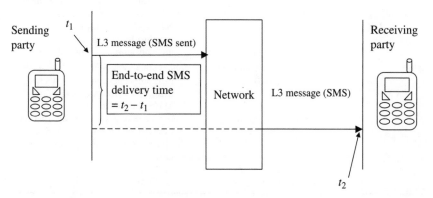

Figure 10-3   End-to-end SMS delivery time.

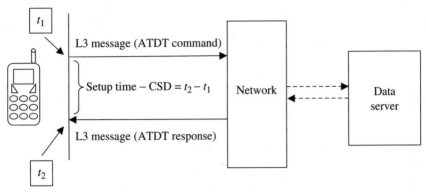

Figure 10-4   Setup time for circuit-switched data call.

TABLE 10-3   GPRS Services and KQIs

Services	Services access—KQIs	Abstract formula
Packet-switched data services (GPRS)	Service accessibility rate	SA = number of successful session attempts/number of session attempts.
	Setup time	See Figure 10-5 for definition.
	Data quality	The end-to-end data quality is measured in terms of average data throughput in both uplink and downlink directions on a best-effort basis. For conversational class data: DQ (uplink) = X bits/second DQ (downlink) = X bits/second For streaming class data: DQ (downlink) = X bits/second
	Completed session ratio	CSR = number of sessions not released other than by end user/number of successful data session attempts.
MMS	Service accessibility rate (send)	Service accessibility MMS-send = number of successful MMS (post) attempts × 100/ number of all MMS (post) attempts.
	Service accessibility rate (retrieve)	Service accessibility MMS-receive = number of successful MMS (get) attempts × 100/number of all MMS (get) attempts.
	MMS send time	See Figure 10-6.
	MMS retrieval time	See Figure 10-7.

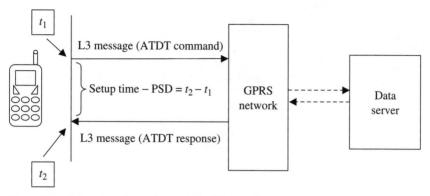

Figure 10-5   Setup time for packet-switched data call.

average throughput in bits/second during a session/call. Depending upon the traffic class, it is measured in uplink, downlink, or both directions. Table 10-4 lists the traffic classes and their characteristics. In addition to average throughput, the minimum and maximum throughput is also measured. To measure the minimum/maximum throughput, the call/session duration is divided into 10 blocks and average throughput is calculated for each block.

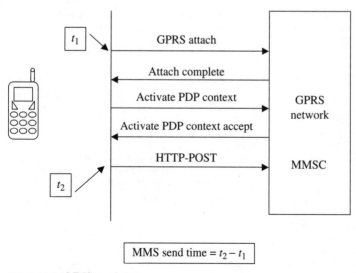

$$MMS \text{ send time} = t_2 - t_1$$

Figure 10-6   MMS send time.

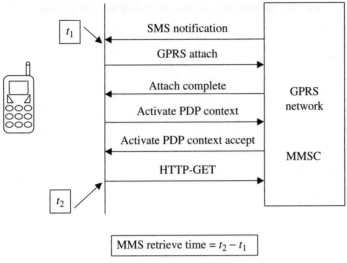

$$\text{MMS retrieve time} = t_2 - t_1$$

Figure 10-7 MMS retrieve time.

TABLE 10-4 Traffic Classes

Traffic class and characteristics	Data quality computation
Conversational class, e.g., voice ▪ Preserve time relation ▪ No buffering ▪ Symmetric traffic ▪ Guaranteed bit rate	DQ (received A side) = X bits/second DQ (received B side) = X bits/second
Streaming class, e.g., streaming video ▪ Preserve time relation ▪ Minimum variable delay ▪ Buffering allowed ▪ Asymmetric traffic ▪ Guaranteed bit rate	Generally, streaming class data quality is measured in the downlink direction. However, if the application requires streaming in the uplink direction, additional measurement may be required. DQ (received A side) = X bits/second
Interactive class, e.g., Web browsing ▪ Request response pattern ▪ Preserve payload content ▪ Variable delay (moderate) ▪ Buffering allowed ▪ Asymmetric bit rate ▪ No guaranteed bit rate	The end-to-end data transmission is computed by the time taken to download a specified file of a fixed data size from a data server. DQ download time [seconds] = $t_2 - t_1$ t_1 = point of time when data request was sent t_2 = point of time when data file was received without any errors
Background, e.g., background download of email ▪ Preserve payload content ▪ Variable delay ▪ Buffering allowed ▪ Asymmetric bit rate ▪ No guaranteed bit rate	Same as in case of interactive class

10.2   Roaming Service Quality

Roaming is a unique service in the sense that the QoS experienced by a roamer is dependent on both the home and the visited networks. For example, once a roamer switches on in a visited network, the update location message is sent by the serving VLR in the visited network to the HLR in the home network, transiting through the international network. It is possible that the update location procedure fails and the roamer is not able to register in the visited network. From a user perspective, the service accessibility is zero. From a technical perspective it may happen, for example, because of a wrong routing table at the serving MSC (visited network), or at a node in the international network, or at the SCCP gateway (home network). Similarly, end-to-end delay in registering in a visited network may be caused by the delay in routing a UL message in the visited network, international network, or home network. This aspect should be taken into consideration while determining roaming KQIs. Figure 10-8 is an attempt to model roaming KQIs. It describes the relationship between the network performance of the home, visited, and international networks with technical QoS.

The method of measurement of the roaming QoS is complex and should be based on both subjective and objective measures. The recommended method to measure user satisfaction in roaming service is to do regular surveys. Many operators have adopted a strategy to proactively monitor the technical QoS indicators to have effective roaming service management. This also helps them to identify service problems early and enable them to take corrective action before the problems result in customer dissatisfaction.

10.3   Proactive Service Monitoring

There are three common performance data sources to derive roaming KQIs, i.e., network elements, results from active tests, and the data from the passive monitoring of roaming transactions. The fault and the performance data from the network elements involved in roaming, such as the HLRs, VLRs, GMSCs, and SCCP gateways, is not service specific but is often used as an indication of the health of the underlying roaming services. One serious limitation here is that the monitoring is limited to the home network. Most of the common solutions available in the market are based on either active tests or a passive monitoring approach. The active approach is based on simulating a roamer, performing tasks exactly in the same way the actual roamer would perform in a real-life scenario. This approach enables an end-to-end service monitoring from an end-customer perspective. The results from this testing are used to derive the service KQIs.

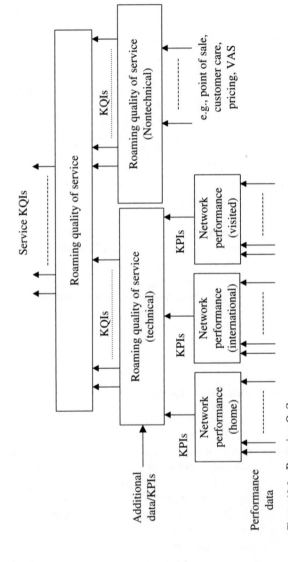

Figure 10-8 Roaming QoS.

The other approach is based on passive monitoring of roaming traffic between partner networks. In this case, 100 percent of the roaming transactions are taken into consideration to determine the service KQIs. Each approach has its own pros and cons.

Figure 10-9 shows the concept of proactive roaming service monitoring using performance indicators from three sources.

10.3.1 Roaming service monitoring using active probes

By simulating roamer activity, this approach is able to measure performance of various services right from the air interface, through the core network, all the way to end applications. The available systems generally support both on-demand and periodic testing. The roaming service performance can be characterized by consolidating results from the periodic testing.

Active probe. GSM MoU PRD IR .41 describes the requirements for QoS measurement equipment. The proposed mobile QoS test equipment (MQT) emulates a typical customer using services offered by the network under test. To simulate user mobility options, it is installed in a fixed location, or in a vehicle, or carried around. The MQT supports both originating and terminating scenarios. More than one MQT is required to test scenarios such as mobile-to-mobile calls and call forwarding. To test mobile-to-fixed-line calls, additional test gear called fixed QoS test (FQT) equipment is required. The FQT also acts as a host for nonvoice call services such as SMS, MMS, WAP, push to talk, and

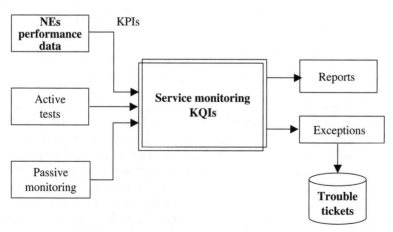

Figure 10-9   Performance monitoring.

Internet browsing. In many implementations the MQT and FQT capabilities are combined in the same test equipment.

As a minimum, the MQT/FQT has the following capabilities:

1. *Necessary hardware to simulate a mobile phone:* A standard mobile phone is controlled by a keypad. The results of the actions are communicated by display. This is not sufficient capability for MQT. The MQT requires flexibility in terms of control, execution, and protocol level information to interpret the results. In most of the implementations available in the market, a high-quality modem is used for the purpose. For MQT, these modems support GSM, GPRS, EDGE, UMTS, and CDMA interfaces. For FQT, they support PSTN and ISDN interfaces. Generally, MQT/FQT supports more than one interface. Usually the modems are connected to the BTS via an antenna.

2. *Remote control capabilities:* The MQTs are expected to be installed remotely and unattended. A capability to remotely reset and initiate the test is required.

3. *Timing and synchronization:* To compute time-based quality indicators, the MQTs are equipped with a good timing source. Moreover, the MQT/FQT simulating the originating side needs to be synchronized with the MQT/FQT at the terminating site. This is achieved either by using same timing server for synchronization or using GPS.

4. *Test functions:* The MQT and FQT support KQI measurements as described in Section 10.1.3. In addition to these most common services used by the roamers, the MQT can also perform tests based on IREG 24, IREG 26, IREG 27, and IREG 35 on a periodic basis.

5. *Test scheduling:* To characterize services over a period, periodic tests are performed.

6. *Data logging:* The test results are logged into a database within a unit or exported to a central unit.

7. *Man-machine interface:* The MMI consists of mechanism to define and schedule the tests and to manage the result data.

Inbound roaming testing. This requires deploying several active probes across the network with SIMs from partner networks. The number of geographically distributed probes depends on the coverage required. Usually, probes are deployed in critical business areas where the roamer concentration is expected to be high. Other possible places are country entry and exit points, e.g., airports. Many of the commercial solutions offer central management of SIMs whereby SIMs stay at a central location and virtual download of SIMs takes place as and when required to remote probes for the test duration. This capability is required as it may not be possible to acquire many SIMs from each partner network to be installed in each remote probe. Depending upon the service type to be tested, one or two probes are required for end-to-end testing.

Figure 10-10 shows a typical architecture for the probe deployment. The system, in general, performs periodic tests as defined in PRD IR.24, IR.26, and IR.27 for GSM, IR.35 for GPRS and IR.52 for MMS. The tests that require manual intervention are excluded. The key quality indicators are then derived from the test results over a period of time.

Outbound roaming testing.   This requires probes to be placed in partner networks. It is not feasible to place probes in all the partner networks because of technical, commercial, and legal issues. The wireless service providers that are operating as part of a global corporation or a business alliance, however, are in a better position to implement outbound roaming testing within their group. In such cases, it is much easier to have cooperation on outbound service monitoring with similar KQIs among the group members. Global roaming test capabilities from independent providers are also commercially available. In this case an independent provider deploys the infrastructure and the wireless service providers lease the test time or use this as a service and pay on the basis of the number of tests performed. The probes are generally placed at strategic locations such as business or financial centers and major tourist spots in a foreign network.

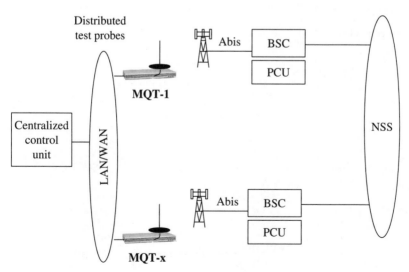

Figure 10-10   Roaming service monitoring—active probe.

10.3.2 Roaming service monitoring using passive probes

Signaling carries rich information about the network, the services, and the subscribers. This information can be extracted by nonintrusive probing of the signaling links. For example, looking at the volume and error rate of the update location procedure to and from a monitored network provides a good indication of roaming service availability from a roamer point of view. This approach offers tremendous benefits as the measured KPIs represent 100 percent roaming traffic rather than a few simulated calls. No additional load is added to the network.

For comprehensive coverage of the roaming services, the links carrying signaling information to and from partner networks are monitored. These links are:

- CCS7 links carrying SCCP-MAP traffic to and from partner networks.

- IP links carrying session management messages to and from partner networks.

Figure 10-11 shows system architecture of a typical monitoring system. It consists of a cardcage consisting of interface and processing cards to extract signaling messages nonintrusively from CCS7 and IP links. The central processor processes the data to measure KPIs.

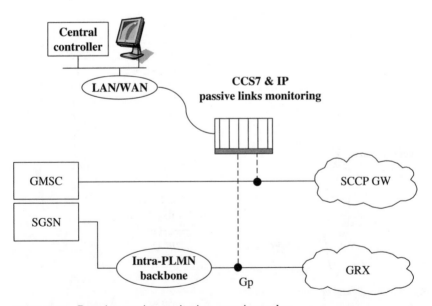

Figure 10-11   Roaming service monitoring—passive probes.

TABLE 10-5   Example Performance Indicators

Services	KPIs
Registration in a visited network	UL success rate UL transaction time (average) GPRS UL success rate GPRS UL transaction time (average)
Security	Send authentication success rate Send authentication transaction time (average)
Roaming number allocation	PRN success rate PRN transaction time (average)
SMS	MO-SMS success rate MO-SMS transaction time (average) MT-SMS success rate MT-SMS transaction time (average)
Session	PDP context activation success rate PDP context activation transaction time

As the monitored traffic is bidirectional, it is possible to generate KPIs for both home and partner networks. The periodic aggregation of volume, errors, and transaction time provides useful data to indicate the quality of roaming service. Table 10-5 lists a few examples of KPIs that can be measured by using passive monitoring of signaling links.

10.4   Troubleshooting Roaming Faults

Roaming is a complex service. The home network, intermediate international carriers and the visited network must all function perfectly under varying conditions, including network load and number and location of roamers, to establish a roaming call. Even though the wireless service providers perform comprehensive tests before service launch to ensure that all the features available to roamers work perfectly, the stability of service is constantly put to the test by dynamic changes in the involved networks. The new software releases, patches, bug fixes, reconfigurations, and routing changes all are prone to errors and may cause roaming service breakdown.

10.4.1   Common network problems

Some of the common problems encountered are as follows. The list is by no means an exhaustive one.

■ Routing table errors

■ E.212 to E.214 translation errors

- Signaling link outage at local end
- Signaling link outage at remote end
- Remote PLMN HLR outage
- MSRN shortage at MSC
- GT not updated
- APN configuration
- Timing issues, timer expires
- SCCP routing problem
- ISUP call routing problem
- Subscriber data in the HLR not correct
- Radio coverage
- Mobile station configuration

10.4.2 Information gathering on symptoms

The very first step to resolve any roaming problem is to gather information on symptoms. It is likely that one may find some pattern to make localizing a fault much easier. Some of the questions one may like to find answers to are as follows.

General

Who is impacted?
- Outbound roamers
- Inbound roamers

What is the problem statement?
- Unable to connect to the network.
- It takes a long time to connect to the network.
- Unable to receive incoming calls.
- Unable to establish outgoing calls to local subscribers in the visited network.
- Unable to establish international outgoing calls.
- Unable to send or receive SMS/MMS.
- Unable to access the Internet in the visited network.
- Unable to access WAP.

Is the problem isolated to one roamer or does it affect a group of roamers?

Is the problem isolated to one specific partner or does it affect a group of roaming partners?

Do the symptoms occur regularly or intermittently? Is there any pattern, e.g., time of day?

Is the symptom repeatable?

Own network

Was there any network reconfiguration in the recent past?

Is there any network migration or expansion happening now?

Is the symptom related to certain network elements, e.g., a subscriber register in one particular HLR, roamers currently visiting a certain geographical area covered by specific MSC/VLR, or SMS submission to a specific SMSC?

Do symptoms point to a certain APN?

Foreign network

Is there any new IR.21 information exchange?

Is my network updated based on latest IR.21 information from the partner networks?

Are the symptoms related to a specific or a group or to all VPLMNs?

10.4.3   Diagnostic tools

Signaling protocol analysis.   Signaling carries rich information on services, network, and subscribers. Monitoring signaling links carrying international roaming traffic often gives valuable clues to localize, diagnose, and possibly resolve faults. One can use stand-alone protocol analyzers, distributed protocol analyzers, or link-monitoring solutions. The link-monitoring solutions offer a great advantage over stand-alone analyzers. A few of the advantages of link-monitoring solutions are as follows.

- Centralized monitoring

- Ability to monitor a few links to several hundred or even a thousand links

- Correlates call legs across the links, e.g., ISUP and MAP correlation

Monitoring points.   The objective of monitoring is to have complete visibility of international SCCP traffic to partner networks. The decision on monitoring points depends on the network configuration. For example, if a wireless operator is also operating an international gateway with SCCP routing support, the links connecting to international carrier/hubs or links directly to partner networks are monitored (Figure 10-12, network

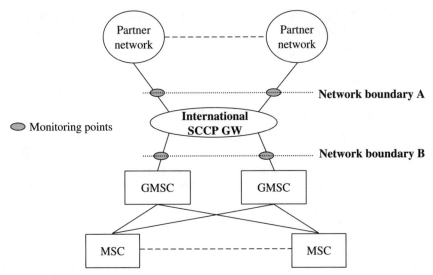

Figure 10-12   Monitoring points.

boundary A). In the other case, where the wireless operator does not have SCCP routing capabilities, the interconnect links from GSMC to the international SCCP gateway are monitored (Figure 10-12, network boundary B).

The links are tapped by using nonintrusive probes. The bridging isolation techniques are used to ensure there will be no impact on the network in case of monitoring equipment failure.

The Gp links from SGSNs to partner networks are also monitored to capture critical information on sessions initiated by inbound roamers. Usually network operators connect to partner networks using services from a global roaming exchange (GRX). In this case the Gp links from SGSN to GRX are monitored.

IREG testers.   IREG testers are the most commonly used equipment for testing and verifying inbound roaming. This equipment is loaded with SIMs from partner networks to simulate inbound roamers. The testing is done from the end-user point of view. The diagnostics information available is very limited.

For outbound roaming testing, a tester requests its counterpart in the partner network to perform the tests and send back the results. The wireless service providers are obliged to perform the tests periodically and on request under the GSM MoU agreement.

10.4.4   Understanding protocol errors

The HPLMN and the VPLMN entities, e.g., HLR and VLR, communicate to each other by using MAP protocol. Understanding of MAP operations, associated errors, and diagnostic information provides greater insight and help in diagnosing, isolating, and resolving roaming faults.

The TCAP provides non-circuit-related information transfer capabilities for a variety of the applications, such as MAP. In the applications in the VPLMN VLR, MSCs use MAP services, which in turn use TCAP to invoke MAP procedures at the HPLMN HLR and other entities. MAP is specifically designed for mobile networks. TCAP relies on the SCCP to deliver signaling messages to other entities across the network. TCAP uses only connectionless services of the SCCP. This means that the SCCP-UDT messages are used only to transport TCAP messages.

Figure 10-13 shows the protocol stack for the communication between the VPLMN and the HPLMN and vice versa. Each of the protocol layers handles errors from its users or providers and takes appropriate action as defined in SCCP, TCAP, and MAP specifications. The following sections describe this aspect in more detail.

A TCAP message consists of two parts, i.e., a transaction sublayer and a component sublayer. The transaction sublayer is responsible for managing the exchange of messages containing components between two TCAP entities. The component sublayer is responsible for component handling between originating and terminating TC users, i.e., HLR, VLR, etc. Figure 10-14 shows transaction and component sublayers and associated message types.

Transaction layer error handling.   The transaction sublayer uses the abort message to terminate a transaction. The transaction is aborted

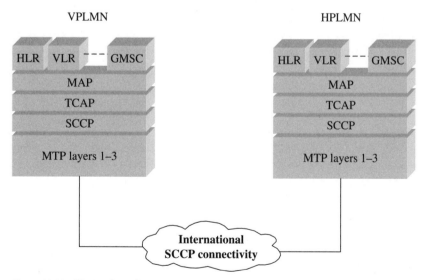

Figure 10-13   Protocol stack.

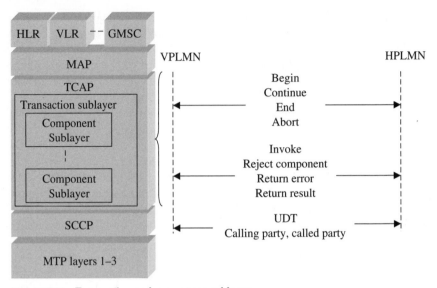

Figure 10-14 Transaction and component sublayer.

following an abnormal condition detected by the transaction sublayer or because of a request by the component sublayer. A reason for terminating the transaction may or may not be given. Two distinct cause codes are provided to identify the source for the termination.

P-abort cause codes are used when the termination request is initiated by the service provider, i.e., TCAP in this case.

U-abort cause codes are used when the user initiates the termination request, i.e., MAP in this case.

Figure 10-15 shows a VLR sending an abort request to an HLR with P-Abort cause code. The transaction is identified by a transaction ID 3b00e8. The TCAP in the MSC/VLR initiated this in response to a continue message received previously with an unrecognized transaction. Table 10-6 shows the provider and user abort causes.

Component portion error handling

Reject component.   If an application is not able to process a component, a reject component is used to convey the nature of the problem to the requester. Table 10-7 lists problem categories and associated problem codes.

```
MT: UDT
  Called party address length: 10 octets
    Subsystem number: HLR
    Translation type: 0
    Nature of address indicator: international number
    Address information: 6596xxxxh
  Calling party address length: 13 octets
    Signalling point code: MSCA
    Subsystem number: VLR
    Translation type: 0
    Nature of address indicator: international number
    Address information: 8613xxxxxxh
  MT: Abort
  Destination transaction ID tag
    Transaction ID: 3b00e8h
  P-Abort cause tag
    P-Abort cause: unrecognized transaction ID
```

Figure 10-15 TCAP abort with P-abort cause code.

Return error component. A return error is sent back if an operation as requested in the invoke component cannot be completed. A return error does not necessarily mean a protocol error. It also includes other causes that prevent an operation from being completed. For example, if a subscriber does not subscribe to roaming services but tries to register in a foreign network, the HPLMN HLR will return an error indicating roaming not allowed. Figure 10-16 shows a subscriber from one of the Singapore networks trying to roam in the Malaysian network. However, VPLMN

TABLE 10-6   Abort and Cause Code

Transaction portion message type	Operation code (hex)	Cause
Abort	67	P-abort ▪ Unrecognized message type ▪ Unrecognized transaction ID/type ▪ Badly formatted transaction portion ▪ Incorrect transaction portion ▪ Resource limitation U-abort The reason is sent within the dialog portion. A common abort cause is application context not supported. This indicates incompatibility in supported protocol version between sender and receiving entity.

TABLE 10-7   Reject Component Problem Codes

Problem code tag	Problem codes	Remarks
General problem	Unrecognized component Mistyped component Badly structured component	
Invoke problem	Duplicate invoke ID Unrecognized operation Mistyped parameter Resource limitation Initiating release Unrecognized linked ID Linked response unexpected Unexpected linked operation	Service not supported Abnormal event detected by peer Response rejected by peer Response rejected by peer
Return result problem	Unrecognized invoke ID Return result unexpected Mistyped parameter	Response rejected by peer Unexpected response from the peer Response rejected by peer
Return error problem	Unrecognized invoke ID Return error unexpected Unrecognized error Unexpected error Mistyped parameter	Response rejected by peer Response rejected by peer Response rejected by peer Response rejected by peer Response rejected by peer

did not allow this subscriber to roam, as the HPLMN HLR returned an error with error code roaming not allowed (see Figure 10-17).

In general, return errors are categorized in following groups:

- Generic errors: System failure, data missing, unexpected data value, etc.

- Identification or numbering errors: Unknown subscriber, unallocated roaming number, etc.

- Subscription errors: Roaming not allowed, illegal services, etc.

- Short message errors: Subscriber busy for MT-SMS, SM delivery failure, etc.

The list of possible MAP return errors associated with each MAP operation code are listed Table 10-8. Understanding of MAP return codes is very helpful in resolving roaming issues. Table 10-9 lists the return errors and their descriptions.

Example scenario

Problem definition.   Inbound roamers from a partner network X, roaming in a business district, are unable to receive incoming calls. However, they are able to make outgoing calls and send/receive SMS.

MT: UDT
   Called Party Address Length: 12 octets
   Subsystem Number: HLR Home Location Register
   Translation Type: Translation Type Not Used
   Nature of Address Indicator: International number
   Address Information: 6598xxxxxxxh
   Calling Party Address Length: 11 octets
   Subsystem Number: VLR Visited Location Register
   Translation Type: Translation Type Not Used
   Nature of Address Indicator: International number
   Address Information: 601xxxxxxxxxh
MT: Begin
Originating Transaction ID Tag
   Transaction Id: 7a60d7d7h
Invoke
Invoke Id Tag
   Invoke Id: 1
Operation Code Tag: Local Operation Code
Operation: Update Location
IMSI Tag
   MCC Digits: 525 MNC Digits: 03 MSIN Digits: 8xxxxxxxxx
MSC Number Tag
   Nature of Address: International number
   ISDN Address Digits: 601xxxxxxxxf
VLR Number Tag
   Nature of Address: International number
   ISDN Address Digits: 601xxxxxxxxf

> Update Location request from a Singapore subscriber currently roaming in Malaysian Network

Figure 10-16   Update location.

MT: UDT
   Called Party Address Length: 11 octets
   Subsystem Number: VLR Visited Location Register
   Translation Type: Translation Type Not Used
   Nature of Address Indicator: International number
   Address Information: 601xxxxxxxxxh
   Calling Party Address Length: 8 octets
   Subsystem Number: HLR Home Location Register
   Translation Type: Translation Type Not Used
   Nature of Address Indicator: International number
   Address Information: 659xxxh
MT: End
Destination Transaction ID Tag
   Transaction Id: 7a60d7d7h
Return Error
Invoke Id Tag
   Invoke Id: 1
Error Code Tag: Local Error Code
   Error Code: Roaming Not Allowed
Roaming Not Allowed Cause Tag
   Roaming not Allowed Cause: PLMN Roaming Not Allowed

> The subscriber is not allowed to roam because of subscription

Figure 10-17   Update location response.

TABLE 10-8   MAP Operation and Return Error Component

Operation	Opcode decimal (hex)	MAP return errors
Update location	2 (02)	System failure Data missing Unexpected data value Unknown subscriber Roaming not allowed
Cancel location	3 (03)	Data missing Unexpected data value
Purge MS	67 (43)	Data missing Unexpected data value Unknown subscriber
Update GPRS location	23 (17)	System failure Unexpected data value Unknown subscriber Roaming not allowed
Provide subscriber info	70 (46)	Data missing Unexpected data value
Send identification	55 (37)	Data missing Unidentified subscriber
Send authentication info	56 (38)	Data missing Unexpected data value System failure Unknown subscriber
Insert subscriber data	7 (07)	Data missing Unexpected data value Unidentified subscriber
Restore data	57 (39)	Data missing Unexpected data value Unknown subscriber System failure
Send routing info for GPRS	24 (18)	Absent subscriber Call barred Data missing Unexpected data value Unknown subscriber System failure
Provide roaming number	4 (04)	Absent subscriber Facility not supported/not allowed Data missing Unexpected data value No roaming number available System failure
Register SS	10 (0a)	Absent subscriber Call barred Data missing Unexpected data value Bearer services not provisioned Teleservices not provisioned

TABLE 10-8   MAP Operation and Return Error Component (*Continued*)

Operation	Opcode decimal (hex)	MAP return errors
		Illegal SS operation
		SS error status
		SS incompatibility
Erase SS	11 (0b)	Call barred
		Data missing
		Unexpected data value
		Bearer services not provisioned
		Teleservices not provisioned
		Illegal SS operation
		System failure
		SS error status
Deactivate SS	13 (0d)	System failure
		Call barred
		Data missing
		Unexpected data value
		Bearer services not provisioned
		Teleservices not provisioned
		Illegal SS operation
		SS error status
		SS subscription violation
		Negative password check
		Number of password attempts violation
Interrogate SS	14 (0e)	System failure
		Call barred
		Data missing
		Unexpected data value
		Bearer services not provisioned
		Teleservices not provisioned
		Illegal SS operation
		SS not available
Process unstructured SS request	59 (3b)	System failure
		Call barred
		Data missing
		Unexpected data value
		Unknown alphabet
		USSD busy
Unstructured SS request	60 (3c)	System failure
		Data missing
		Absent subscriber
		Unexpected data value
		Unknown alphabet
		USSD busy
		Illegal subscriber
		Illegal equipment
Unstructured SS notify	61 (3d)	System failure
		Call barred
		Absent subscriber
		Unexpected data value

(*Continued*)

TABLE 10-8   MAP Operation and Return Error Component (*Continued*)

Operation	Opcode decimal (hex)	MAP return errors
		Unknown alphabet USSD busy Illegal subscriber Illegal equipment
Send routing info for SM	45 (2d)	System failure Call barred Data missing Unexpected data value Teleservices not provisioned Facility not supported Unknown subscriber Absent subscriber SM
MO forward SM	46 (2e)	System failure Unexpected data value Facility not supported SM delivery failure
MT forward SM	44 (2c)	System failure Unidentified subscriber Data missing Unexpected data value Facility not supported Unknown subscriber Absent subscriber SM Illegal subscriber Subscriber busy for MT-SMS SM delivery failure
Report SM delivery status	47 (2f)	Data missing Unexpected data value Unknown subscriber Message waiting list full

Analysis.   By analyzing problem statement it is clear that either the provide roaming number procedure is not successful or incoming calls are not routed correctly.

Diagnostic.   The first step is to isolate the fault between the roaming procedure and the ISUP call routing. A protocol session may help in this case. The recommended steps are as follows.

1. Select signaling links carrying traffic to partner network X.

2. Set up an appropriate filter to reduce the amount of captured traffic. This is required for efficiency purposes and to focus on the problem in hand.

   ■ SCCP MSUs only.

   ■ SCCP calling party address partner network X. For example, if the partner network is Vodafone, then it could be set to +4412-, where

TABLE 10-9   MAP Error Codes

MAP errors	Error code decimal (hex)	Brief description
Unknown subscriber	1 (01)	No subscription exists.
Unknown MSC	3 (03)	
Unknown location area	4 (04)	
Unidentified subscriber	5 (05)	The database (HLR/VLR) does not contain any entry for this subscriber. It is not possible to determine whether the subscription exists.
Absent subscriber SM	6 (06)	MT-SMS transfer cannot be completed because network cannot contact the MS.
Unknown equipment	7 (07)	
Roaming not allowed	8 (08)	The user is not allowed to roam in an area because of subscription.
Illegal subscriber	9 (09)	The subscriber is not allowed to access services, as authentication failed.
Bearer service not provisioned	10 (a)	
Teleservices not provisioned	11 (b)	
Illegal equipment	12 (c)	IMEI check procedure shows that MS is not white-listed.
Call barred	13 (d)	
Forwarding violation	14 (e)	
CUG reject	15 (f)	
Illegal SS operation	16 (10)	
SS error status	17 (11)	
SS not available	18 (12)	
SS subscription violation	19 (13)	
SS incompatibility	20 (14)	
Facility not supported	21 (15)	The PLMN/terminal does not support the requested facility.
Invalid target base station	23 (17)	
No radio resources available	24 (18)	
No handover number available	25 (19)	
Subsequent handover failure	26 (1a)	
Absent subscriber	27 (1b)	
Incompatible terminal	28 (1c)	
Short-term denial	29 (1d)	
Long-term denial	30 (1e)	
Subscriber busy for MT-SMS	31 (1f)	MT-SMS transfer cannot be completed because another MT-SMS transfer is going on.
SM delivery failure	32 (20)	
Message waiting list full	33 (21)	
System failure	34 (22)	The requested task cannot be completed because of a problem in another entity. The type of resource or entity may be given in the resource indicator parameter.
Data missing	35 (23)	An optional parameter required by the context is missing.

(Continued)

TABLE 10-9   MAP Error Codes *(Continued)*

MAP errors	Error code decimal (hex)	Brief description
Unexpected Data Value	36 (24)	The data type is valid as per specifications but its value or presence is unexpected in the current context.
PW registration failure	37 (25)	
Negative PW check	38 (26)	
No roaming number available	39 (27)	A roaming number cannot be allocated because all available numbers are in use.
Tracing buffer full	40 (28)	
	41 (29)	
Target cell outside group area	42 (2a)	
Number of PW attempt violations	43 (2b)	
Number changed	44 (2c)	The subscription does not exist for that number anymore.
Busy subscriber	45 (2d)	
No subscriber reply	46 (2e)	
Forwarding failed	47 (2f)	
OR not allowed	48 (30)	
ATI not allowed	49 (31)	Any time interrogation.
No group call number available	50 (32)	
Resource limitation	51 (33)	
Unauthorized requesting network	52 (34)	
Unauthorized LCS client	53 (35)	
Position method failure	54 (36)	
Unknown or unreachable LCS client	58 (3a)	
MM event not supported	59 (3b)	
ATSI not allowed	60 (3c)	Any time information handling.
ATM not allowed	61 (3d)	
Information not available	62 (3e)	
Unknown alphabet	71 (47)	
User busy	72 (48)	

"-" is a wild character. The wild character allows capturing of all MSUs with country code 44 and network code 12; the rest of the digits are insignificant. Note that protocol analyzers from different vendors support different wild characters.

- SCCP called party address MSC serving business district.

3. Capture the traffic for a substantial time, say 15 minutes or more.

4. Stop the protocol analysis session and analyze the return errors. Generate statistics on error distribution if PA supports this feature.

5. In this example (see Figures 10-18 and 10-19), most of the PRN transactions are returned with an error. The error type shows no roaming
number available. On further analyzing the traffic by looking at
transactions with errors, it is evident that the MSC is running out
of roaming numbers.

Problem resolution. The next step is to resolve the problem. Further
analysis is required to establish if this problem occurs all the time or
occasionally. If it occurs occasionally, what is the pattern, which day of
the week, which hour of the day, and so on. Once the facts are established, an appropriate action is taken. For example if the provide roaming number procedure is failing consistently with the return error no
roaming number available, it is likely that the MSRN range assigned
to the MSC/VLR is not sufficient. The problem can be resolved by adding
number blocks to the existing MSRN range.

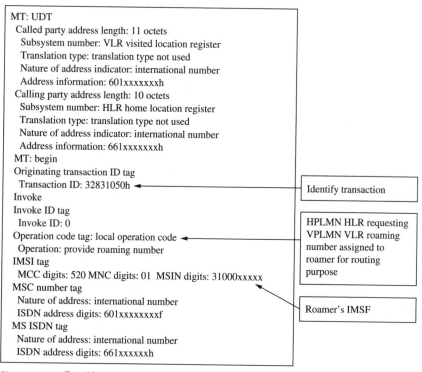

Figure 10-18   Provide roaming number protocol decodes.

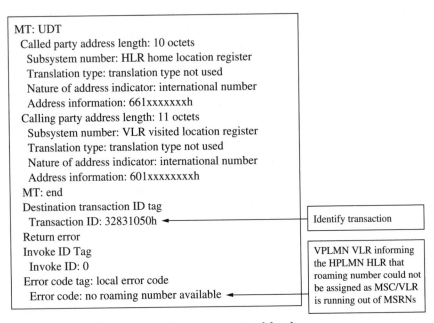

Figure 10-19   Provide roaming number response protocol decodes.

Bibliography

ITU-T E800, Terms and definitions related to Quality of Service and Network performance.

ETSI ETR 003, Network Aspects (NA), General aspects of Quality of Services(QoS) and Network Performance (NP).

GSM MoU Association PRD IR.41, Identification of Quality of Service aspects of popular service.

GSM MoU Association PRD IR.42, Definition of Quality of Service parameters and their computation.

GSM MoU Association PRD IR.43, Typical procedures for QoS measurement equipment.

GSM MoU Association PRD IR.44, Requirement for QoS measurement equipment.

3GPP TS 29.002, Mobile Application Part (MAP) specification.

Chapter

11

Billing and Settlement

International roaming allows a subscriber to access services virtually anywhere in the world. The visited network obviously needs to charge foreign subscribers for access time, transport, and services. As the visited network is not in a position to directly bill the roamers, it invoices their home network for the service usage. The home network then charges its own subscribers for the services used while roaming in a foreign network, using standard retail billing mechanisms. Figure 11-1 shows the interoperator billing process in its simplest form. John, a subscriber from PLMN A, roams in PLMN B. PLMN B allows John to use the services and collects the usage data. PLMN B then sends invoices and detailed records for service usage to John's home PLMN, i.e., PLMN A. PLMN A settles the invoices according to the roaming agreement. PLMN A also generates a retail bill and charges John for the usage while roaming in PLMN B.

11.1 Roaming Billing Standards

The number of roaming partners spread over different continents makes the interoperator billing process quite challenging. Moreover, complexity is added by partners using different network element technologies with varying billing data formats and transport methodologies. With the introduction of data roaming with GPRS and 3G and associated new services, the billing and settlement among wireless service providers is even more complicated. Several international industry forums and standards bodies are working to simplify and standardize this process for wireless service providers. Two of the GSM Association working groups, i.e., Billing, Accounting, and Roaming Group (BARG) and Transfer Accounting Data Interchange Group (TADIG) have contributed significantly in terms of standards and charging and billing principles for

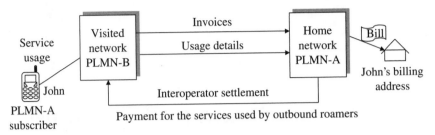

Figure 11-1   Interoperator billing and settlement process.

wireless operators using GSM technologies. The CDMA Development Group (CDG) has developed processes and specifications for wireless operators using CDMA technologies.

TAP (Transferred Account Procedure) is the file format defined by GSM MoU TADIG and is used by GSM operators to exchange billing information with partner networks. The information includes both the charges that each roamer has to pay for the service usage as well as the detailed information related to the services used.

CIBER (Cellular Intercarrier Billing Exchange Roamer) is a proprietary protocol and specification for the exchange of roaming billing information, for voice and data. The wireless service providers using AMPS-based technologies, such as analog AMPS, IS-136 TDMA, and IS-95 CDMA, use CIBER records.

11.2   Transferred Account Procedures

Transferred Account Procedure is the mechanism by which wireless operators exchange roaming billing information. The GSM Association in 1995 released the very first TAP specification, i.e., TAP version 1. From then onward, the specifications have continuously evolved to support new services and associated operational aspects. For example, in the earlier versions, the usage was recorded in terms of call detailed records (CDRs). This was changed to call event detail (CED) to reflect the nature of services in current and future generations of networks. TAP2 and TAP2+ were introduced later; they allowed the operators to bill for new services and provide the additional information required by satellite and U.S. operators. TAP3 is the latest version released by the GSM Association. It includes all the features supported by earlier versions of TAP. In addition, it supports billing for new-generation services, including mobile multimedia and prepaid roaming. It also supports interoperator tariff (IOT) charging principles and key information for marketing and customer service functions.

11.2.1 TAP3 and earlier versions

The latest version, TAP3, is designed to cater to the needs of the next-generation services. It uses an industry-standard coding scheme, i.e., ASN.1. This enables use of commercially available tools rather than proprietary toolkits. Many of the earlier version constraints, e.g., file size limitation, have been also removed.

TAP3 can handle all the features and services supported by earlier versions of TAP. In addition, the new supported services are:

- High-speed circuit-switched (HSCSD) and packet-switched (GPRS) data services

- Prepaid roaming using CAMEL

- USSD charging

- Mobile directory number to support mobile number portability for interstandard roaming

- Support of private numbering plans (SPNP)

- Enhanced full rate (EFR) for enhanced voice quality

- Fraud information gathering system (FIGS), including fraud monitoring indicator and third-party number

- Support for UMTS QoS

The additional interoperator tariff features that TAP3 is able to handle are:

- HPLMN repricing—enables HPLMN to reprice each call according to its own tariff plan

- Call-level discounts

TAP3 also allows for the specific requirements of satellite networks and of large countries where no single operator covers the entire geographic area. This includes support for the following additional parameters.

- Additional charging parameters, e.g., separation of airtime and toll charges

- Additional time zones

Unlike earlier versions, TAP3 also contains valuable information about roamer and also services used by a roamer. Wireless service providers can use this information for marketing (e.g., targeted campaigns) and customer care. This information could also be used to build roamer profiles and ad hoc studies as and when required. This enables various stakeholders within a roaming organization to make informed business decisions.

11.2.2 TAP-in and TAP-out processes

In the GSM world, the usage records are generated in mobile switching centers (MSCs), short message service centers (SMSCs), and voice mail service centers (VMSCs). Several different types of records are generated, depending on the usage. For example, call detailed records (CDRs) for mobile-originated (MO) and mobile-terminating (MT) calls, transaction detailed records for MO-SMS, MT-SMS, and other nonvoice usage. In GPRS and 3G, usage records are generated at the SGSNs, GGSNs, MMSCs, and a host of other gateway elements. The usage records are generated for packet-switched data calls, MMS, and access of contents.

The TAP-out process enables an HPMN (the PLMN where TAP-out processing is performed) to send rated records for the calls made by inbound roamers (visitors from foreign networks) to their respective home networks (VPMNs).

As shown in Figure 11-2, the serving MSC in the visited network creates detailed records every time a roamer successfully accesses a service. These records are then transferred to the billing system for rating and pricing. The billing system segregates and group calls/records created for the roamers and converts those in the ASN.1 TAP file format. The MCC and MNC codes in IMSIs are used to validate and group the calls. The TAP files contain rated call information. The rating is done in accordance with the bilateral agreement between operators. These TAP files are then sent to roaming partners on a regular basis. This transfer takes place either directly or via a clearinghouse. The frequency of transfer is subject to the bilateral roaming agreement and generally decided up-front. In general, the TAP file exchange should take place as frequently as possible to enable monitoring of high usage and possible fraud.

Electronic data interchange (EDI) is the standard mechanism of TAP file transfer to ensure that charging records are made available to the HPLMN without delay. In case of EDI failure, magnetic tapes or some other suitable mechanism can be used as a fallback and are subject to a bilateral agreement. Magnetic tape technology is fast becoming obsolete.

The TAP-in process at the receiving network accepts the files generated by its partner networks. The TAP-in process involves parsing, validation, conversion of usage data into internal format, and prerating in accordance with the roaming tariff plan. The reject and return process is used in the cases where the validation of TAP files results in errors.

11.2.3 Reject and return process

A new procedure called *reject and return* was introduced recently as part of the TAP3 specification in order to handle errors in TAP files efficiently. Before the implementation of this process, an error concerning one single call in a TAP file resulted in rejection of the entire file. This was the cause of unnecessary delays in the billing and settlement process.

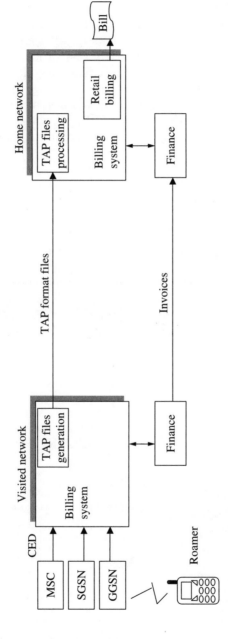

Figure 11-2 Transferred account procedure.

The reject and return process allows processing of validated CEDs to proceed and return of errored CEDs back to the concerned VPMN. An automated mechanism can be built to handle the fatal errors and missing files or data. Having fewer call event details at the end of the month in the reclaims process allows an early interoperator settlement.

Having the capability to reject individual call event details also benefits the retail billing process and early realization of dues from subscribers. Figure 11-3 describes a simplified view of TAP file transfer using the rejects and returns process.

At the HPLMN, this process enables the return of call event detail records containing severe errors to a concerned VPLMN, while correct CEDs can be processed as usual. The typical subprocesses at the HPLMN are:

- TAP file validation
  - Missing file detection
  - Fatal error detection
  - Severe error detection
- Creation of a RAP file
- Transmission of RAP files to the concerned VPLMN

The VPLMN, if possible, corrects the files and/or call event details and resubmits them to the HPLMN. This process allows for the VPLMN to recoup roaming revenues from the HPLMN for resubmitted call event details/files. The typical subprocesses at the HPLMN are:

- RAP file decoding
- Submit missing file/files
- Correct fatal errors
- Correct severe errors
- Create TAP files and resubmit to the HPLMN

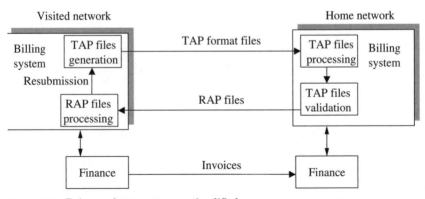

Figure 11-3   Reject and return process simplified.

11.3   Role of Clearinghouses

Wireless networks are a heterogeneous group, being based on different technologies and standards. Each operator needs to partner with many networks (typically 200 or more) to meet its subscribers' roaming needs. This means that wireless service providers need to manage multiple relationships, interconnect globally, and exchange data with multiple formats. On the commercial side, they need to manage complicated financial relationships with varying laws and regulations in different parts of the world. Figure 11-4 shows the three possible relationship scenarios between operators.

In scenario (a), the operators have a direct relationship. This is a preferred option for the operators in neighboring countries where heavy

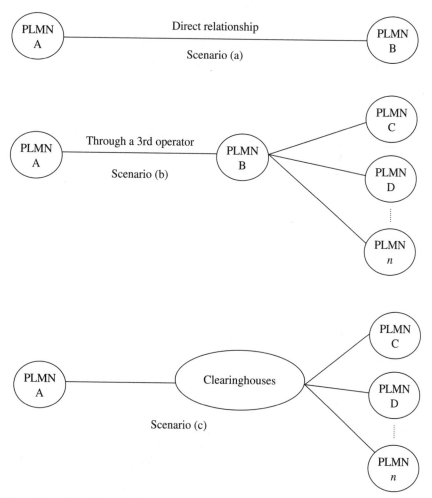

Figure 11- 4   Roaming relationship scenarios.

roaming traffic is expected. It is also used between operators having the same management. Operators offering national roaming also commonly deploy this scenario. Although wireless service providers do have a choice to connect directly to all their partners, it is obvious that a point-to-point relationship with each roaming partner is not feasible and is cost-prohibitive.

Scenario (b), often called a piggyback relationship, is used between operators that are members of the same alliance or managed by the same group.

In scenario (c), an independent clearinghouse is used. Clearinghouses play a crucial role in making a global roaming footprint a reality. The primary aim of clearinghouses is to support wireless service providers in managing their global roaming operations. For new network operators, the clearinghouse helps them to speed up market entry. In general, clearinghouses offer the following services or subset of services:

- *Roaming agreement broker.* The clearinghouse takes full responsibility for negotiation, preparation, and documentation of a roaming agreement on behalf of a PLMN.

- *Roaming partner management.* The clearinghouse acts as a single point of contact for managing multiple roaming partners.

- *Facilitation of interstandard roaming.* This includes, for example, management of different settlement and billing formats, signaling standards, switch incompatibilities, and proprietary extensions and of other technology issues.

- *Data clearing and settlement.* The clearinghouse provides a centralized process for the exchange of TAP/CIBER data, monthly financial settlement, electronic data delivery, management of IOT and currency, etc.

- *TAP conversion.* To support industry needs, GSMA releases several TAP version releases every year. Most recently, it has decided to release two versions annually. This means that wireless service providers need to upgrade their billing systems to support new TAP releases all the time. The frequent billing system upgrades are not desirable, considering roll-out time and cost. Many service providers prefer to maintain a certain TAP version and outsource conversion to the latest TAP version to clearinghouses.

- *Customer care and business intelligence.* The clearinghouse acts as source for roaming activity. For example, with the data they have, they can easily build roamers' profiles, e.g., where they come from, how long they stay, and what services they use. This is extremely useful information for customer care and marketing.

- *Fraud management.* The clearinghouse helps operators in detecting frauds early. Fraud types such as cloning and subscription can be handled effectively. The TAP3 specification calls for high-usage reports to be sent to HPMN to combat roaming fraud. The VPMN can implement a fraud management system on its own or outsource this function to clearinghouses to meet its obligation.

In addition, some clearinghouses also provide CCS7 and IP connection paths between partners.

Bibliography

PRD TD.57, Transferred Account Procedure Data Record Format Specification, Version Number 3.
www.gsmworld.com.

Roaming Value-Added Services

International roaming offers a great opportunity for wireless service providers to improve revenues and generate more profits. They are continuously exploring new ways to attract more roamers to their networks. Generally, roamers in a visited network register automatically, as most mobile phones are set to automatic network selection ON as a default. Mobile phones in the automatic network selection mode choose a network with the best signal strength at that particular point and place. Reselection may also happen automatically as a roamer moves from one area to another. In general, roamers are not loyal to any particular network while visiting outside their own network domain. They may switch to any network if they like, and if there is any reason to do so. The question is how to attract them to manually select a particular network.

The cost of usage, services, discounts, etc. may surely attract more roamers to switch to a particular network. A promotion has to be carefully planned to offset revenue losses due to discounts with the acquisition of additional roamers. The majority of roamers are business users, and they may not really worry too much about the cost. This approach, therefore, has its own limitations.

The brand awareness, strategic relationship with partners, and alliances with service providers serving different geographic areas are influencing factors too. Aggressive and sound marketing strategies are required to build an image to gain roamers' favor.

The quality of service is a very important factor to influence a roamer to select a particular network. Most roamers are business customers; they rely on roaming services to conduct their businesses. They are very particular about QoS offered and switch easily, if not satisfied.

Another important factor that may influence roamers to be loyal to a particular network are the value-added services offered. This section focuses on the common value-added services generally offered by many wireless service providers. The services listed in this chapter are not exhaustive.

12.1 Optimal Routing

Figure 12-1 shows two PLMNs, i.e., PLMN A and PLMN C in country x and PLMN C in country y. When a PLMN A subscriber (MS A) dials to call a subscriber in PLMN B (MS B), the originating switch analyzes the dialed digits and routes the call to PLMN B using the international network. The GMSC at PLMN B queries the HLR for routing information. The HLR checks for subscriber data to identify the current serving MSC/VLR, which is a VLR in PLMN C in this example. The HLR requests the VLR to send a routing number. The VLR assigns a routing number and sends it to the HLR in a response message. The HLR passes back the MSRN to the GMSC, which routes the call to PLMN C using an international network. This is surely not an efficient routing, as two international call legs are required even though both calling and called parties are currently in the same country. The calling party pays international call charges for the originating leg and the called party pays for the international terminating leg plus a roaming surcharge.

Many wireless service providers are implementing optimal routing (OR) as a value-added service to attract and retain roamers in their network. Optimal routing allows a call to be routed directly from the originating MSC to the MSC currently serving the roamer. This way both the international call legs can be avoided, saving cost. Part of the cost saving can be passed to the roamer and the caller. Optimal routing is also referred to as local direct dial (LDD) because it reduces international call legs to a single local call.

Depending upon implementation, optimal routing can be applied to the following call scenarios.

- Inbound roamer calling another inbound roamer, currently roaming in the same network.

- Inbound roamer calling another inbound roamer, currently roaming in a different network but in same country.

- A local subscriber calling an inbound roamer in the same network.

12.1.1 Implementation

Wireless service providers educate their own subscribers and inbound roamers about optimal routing, its benefits, and methods to invoke this service. For the local subscribers, a normal marketing mechanism is used. The only way to inform inbound roamers is to send SMS on their arrival or registration in the PLMN or by advertising at entry and other

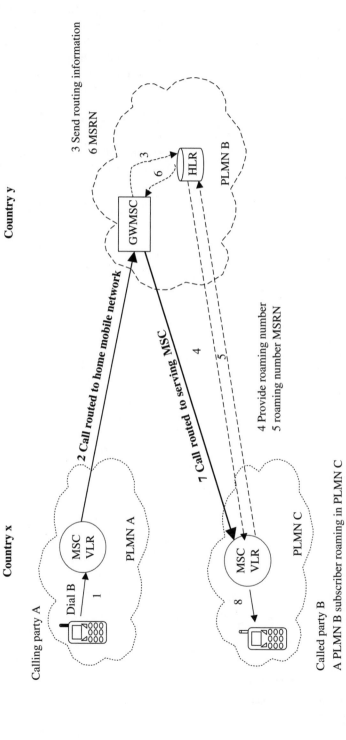

Country y

3 Send routing information
6 MSRN

PLMN B

2 Call routed to home mobile network

1 Call routed to serving MSC

4 Provide roaming number
5 roaming number MSRN

Country x

Calling party A

Dial B

PLMN A

PLMN C

Called party B

A PLMN B subscriber roaming in PLMN C

Figure 12-1 Call routing to a roamer.

prominent places. Usually, to invoke this service, an access code is used followed by the regular B party international number. All the OR/LDD calls are then routed to a special platform (let us call it the OR platform for explanation purposes), which enables optimal routing.

There could be several different ways to implement OR in the network. The three important requirements for any OR implementation are:

1. Acquire, maintain, and manage the data for all inbound roamers currently in the network in real time.

2. Analysis of the B party number to check if it is an inbound roamer currently in a network implementing OR or in any other PLMN within the same country, where the call could be routed locally.

3. If OR is possible, then acquire MSRN from the serving MSC to route the call locally.

One of the efficient ways to build roamers' information in real time is to monitor MAP transactions between the HPLMN and VPLMNs. Figure 12-2 shows a typical implementation of a probe-based solution to

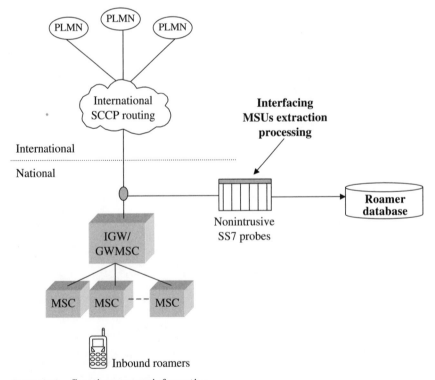

Figure 12-2 Creating roamer information.

monitor the transaction nonintrusively. It consists of acquisition hardware and the necessary processing capabilities to extract the signaling messages from the CCS7 links carrying MAP traffic to partner networks.

The key advantages of this overlay solution approach are:

- It is nonintrusive
- No additional processing is required on the network side
- It operates in real time
- It is independent of make and version of the network elements involved
- It is scalable

The key roaming procedures, such as update location, insert subscriber data, authentication, and cancel location messages provide all the critical information to build the database for OR purposes. For example, by decoding a UL/ISD procedure, the following parameters can be extracted.

- Subscriber's IMSI
- MSISDN
- HLR address
- Serving MSC/VLR address
- Timestamp for first and subsequent UL

Figure 12-3 shows a typical implementation. A fixed line subscriber, a local mobile subscriber, or a roamer dials the OR access code followed by a B party number. The B party number is the international number of a roamer currently roaming in PLMN B. The local originating switch or originating MSC analyzes the OR access code and passes the ISUP IAM to the OR platform. The OR platform checks if the B party is currently roaming in PLMN B by querying the roamer database. If no, the OR platform releases the call after feeding an appropriate announcement to the calling party or processes the call in the normal nonoptimal way. If yes, it extracts the roamer information from the database, using MSISDN as the key. The information includes roamer IMSI, last known VLR, etc.

The OR platform then invokes the MAP provide roaming number procedure with the serving VLR to get the MSRN. Once the MSRN is available, the OR platform routes the call to the serving MSC.

In the above implementation, it is not possible to route the call if the roamer moves to another PLMN in the same country. Figure 12-4 shows a minor variation to overcome this problem. In this case, the OR platform invokes the MAP send routing information (SRI) procedure with the HPLMN HLR to get routing information. The HLR, on receiving the SRI

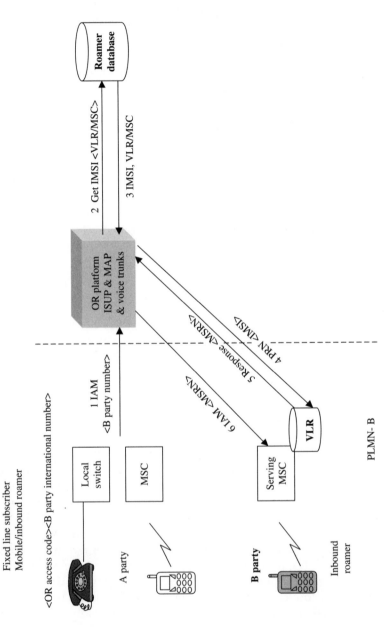

Figure 12-3 Optimal routing implementation 1.

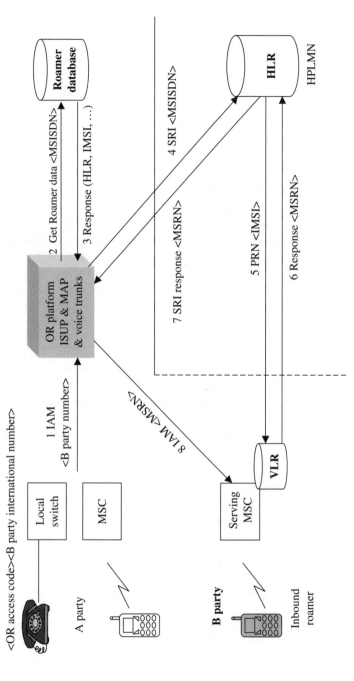

Figure 12-4 Optimal routing implementation 2.

request, invokes the MAP provide number procedure with the serving VLR. The serving VLR assigns a temporary identity to the roamer, i.e., MSRN, for call routing purposes and sends the assigned MSRN to the HLR in its response message. On receiving the MSRN, the OR platform analyzes to check if the roamer is currently in another PLMN within the same country. If yes, it routes the call as a local interconnect call.

The typical implementations discussed in this section are just for illustration purposes. No attempt is made to cover all the functionalities required to build a system for commercial purposes. More intelligent solutions can be deployed by using IN and CAMEL capabilities.

12.2   Welcome and Other Information SMS

Many operators have implemented this service as a means to remain in touch with their roamers and visitors. This service allows wireless service providers to send custom messages to roamers and visitors on first registration and/or at a later time. Typically, the content of the SMS is customized on the basis of the roamer profile. For example, a welcome SMS could be sent in the roamer's native language and in a certain format to meet cultural demands of the region the roamer came from. Subsequent informative or promotional messages are also delivered periodically or on a one time basis. How much a roamer values welcome and subsequent messages is a big question mark. A regular roamer may find it very awkward to receive multiple SMSs, while they could be a great help to a new roamer. Information like currency rates, emergency numbers, weather information, and hotel and taxi booking numbers could be of great help to a visitor in a foreign network. The best way to implement this service is to provide a mechanism by which a roamer can block the message if desired.

12.2.1   Implementation

The essential elements for any implementation are:

1. Acquire, maintain, and manage the data for all inbound/outbound roamers in real time.

2. Create customized SMSs dynamically.

3. Provide an SMSC interface for SMS submission.

Figure 12-5 shows a typical implementation of such a system. It is not our intention to provide a complete system design. This diagram is included for just reference purposes and to help develop understanding.

CCS7 link monitoring or any other suitable mechanism could be used to get roamer information on registration. For a CCS7-based

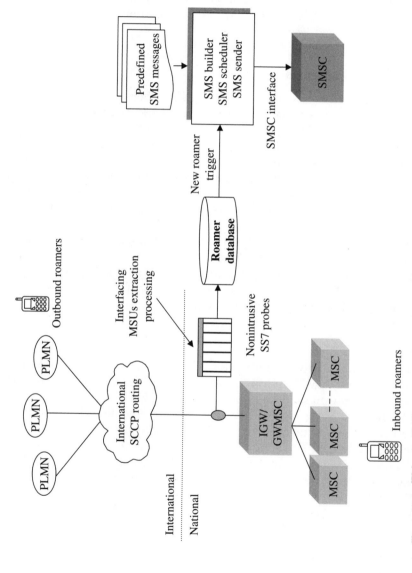

Figure 12-5 Welcome SMS system—an example implementation.

implementation, update location messages to and from partner networks are analyzed to get the information on new roamer registration
in the HPLMN or any of the VPLMNs. On receiving a new roamer trigger, the system selects predefined messages from the message catalog,
codes it in SMS format, and submits it to SMSC using standard interface. SMSC then delivers the SMSs to inbound or outbound roamers
using usual procedures.

12.3   Short Dial Codes

Roamers are generally accustomed to using short codes for accessing
value-added services while in their home network. Many of them tend to
use the same short dial codes while roaming in foreign networks because
of lack of awareness, resulting in call failure. For example, the short code
for voice mail in one of the PLMNs is 1313. However, while roaming, one
needs to dial a complete international number, i.e., +601330001313 and
not 1313. The visited network has no clue to short codes used in other
PLMNs and will not be able to route calls to a value-added service platform at the HPLMN.

Many wireless service providers have implemented short dial code services for their inbound roamers. This enables roamers to access the services transparently, in exactly the same manner as if they were in their
home network. The wireless service providers benefit by increased usage
of home-based services and better completion rates.

12.3.1   Implementation

There are many ways to implement this service. One of the simplest
ways is for the PLMN to maintain the database for short codes and corresponding international numbers for each partner network.

As shown in Figure 12-6, when an inbound roamer dials a home-based
short code, the serving MSC routes the call to the short code platform for
further processing. The short code platform analyzes the dialed code in
relation to the roamer's country and network codes and translates the short
code to the international number, using a database. The MSC then routes
the international call in the normal way to the roamer's HPLMN, using
the international network.

12.4   Missed-Call Information While
Roaming

A roamer in a visited network may miss an incoming call. It may be inten-
tional, as the roamer may choose not to answer for any of a variety of reasons; for example the roamer may be busy in a meeting or may avoid

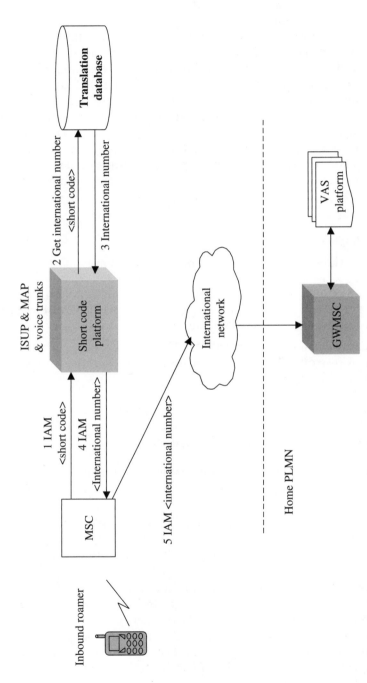

Figure 12-6   Short dial codes—an example of implementation.

incoming calls from an unknown caller to save cost. It may also be unintentional, as the roamer may be away from the phone or the phone may be switched off. In any case, it is desirable to get caller ID, as in the case of calls in the home network. Unlike in the home network, calling line identity (CLI) does not work while the mobile phone is roaming. Many wireless service providers have implemented missed-called alerts for their outbound roamers if the call originates from the home network. These alerts are sent via an SMS containing caller ID. This can be offered as a subscription service. Charging a subscription fee and the increased calls to the home network benefit the wireless service provider. With this feature in place, roamers have knowledge of all the calls from the home network. Roamers can also use this feature to screen the incoming calls and call back to important numbers only.

12.4.1 Implementation

The essential elements to implement this service are:

1. Real-time monitoring of MAP provide roaming number procedures toward partner networks. This provides correlation of IMSI in the forward message with the MSRN in the response message.
2. Real-time of monitoring of ISUP IAM and REL messages to roamers. This filters all the calls to the roamer with a specific release cause code, e.g., busy or no answer.

   Figure 12-7 shows one of the possible implementations.
   The CCS7 SCCP links carrying MAP traffic to partner networks are monitored for the following MAP procedures.

- *Update location:* This will provide critical information such as roamer IMSI, MSISDN, and country and network where the roamer is currently registered.

- *Provide roaming number:* This indicates that the HPLMN needs to terminate a call to its outbound roamer. The IMSI, MSISDN, and MSRN correlation can be built by using UL, PRN, and PRN response.

The system monitors the ISUP links for all the calls routed to a partner network with the same B party number that the MSRN previously captured in the PRN procedure. If the call gets completed, then no further action is required. If the call failed for reasons specified by the administrator (e.g., user busy, no answer), then a trigger is set to send an SMS informing the roamer of the CLI as extracted from the IAM message, i.e., calling party number.

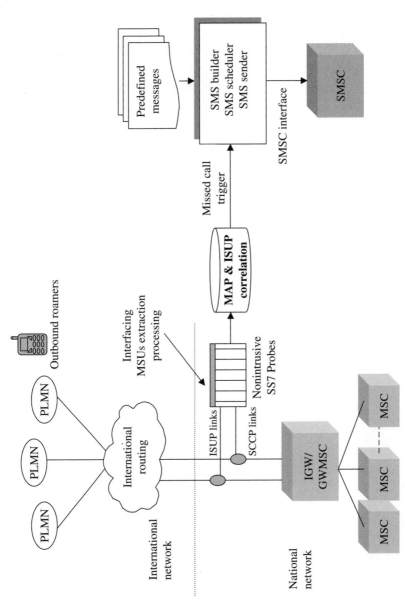

Figure 12-7 Missed call alert—example implementation.

12.5   Other Services

In addition to services discussed in previous sections, service providers have deployed many other value-added services. A few examples of such services are:

- Automatic dialing error correction
- International voice VPN
- GPRS data roaming services: WAP, Internet access, Blackberry, MMS, etc.

13

PLMN—WLAN Roaming

13.1 WLAN Overview

Wireless local area networks (WLANs), also referred as Wi-Fi, have been around for many years. The initial WLAN deployments were restricted because of proprietary protocols. Later, IEEE standardized the WLAN in its Specification 802.11. Presently, there are three main variants available, i.e., 802.11a, 802.11b, and 802.11g. WLANs based on 802.11b are the most popular and widely used in hot spots and personal devices such as PDAs and laptops. WLANs based on 802.11b support data rates up to 11 Mbps. WLANs based on 802.11a and 802.11g can support data rates up to 54 Mbps. WLANs based on 802.11g are backward compatible with 802.11b. This means that the 802.11b devices work in a coverage area of 802.11g-based WLAN.

WLAN consists of two main components, i.e., a station and an access point.

Any device that contains a MAC (medium access control) and Physical Layer based on IEEE 802.11 is termed a station (STA). The main functions of a station include authentication, privacy, and the MAC service data unit (MSDU) delivery.

An access point (AP) is a stationary device that acts like a base station in a wireless network. It allows access and communication between stations. Moreover, it also acts as a bridge between the WLAN and the wired network, if required. Generally, APs are implemented as a stand-alone devices. Other variations are also available, e.g., a personal computer equipped with the wireless network interface card. Access points contain all the functions of a station plus other functions including security, association, distribution, and integration.

Figure 13-1 shows a basic service set (BSS) and network topologies for a WLAN. In its simplest form, two or more stations communicating

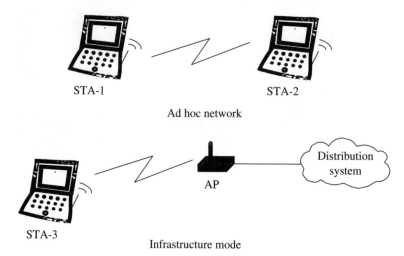

STA-1   STA-2

Ad hoc network

AP

Distribution system

STA-3

Infrastructure mode

Figure 13-1   Basic service set.

with each other form an ad hoc WLAN. This topology is called an independent basic service set (Id-BSS). There is no access possible to the wired network in the ad hoc mode; it is limited to peer-to-peer communication only.

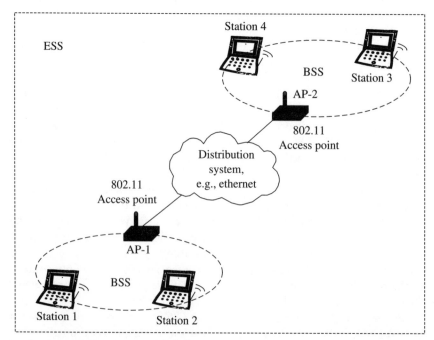

Figure 13-2   Wireless LAN architecture.

In the infrastructure mode, an AP is used to facilitate exchange of data within the BSS. A BSS with this topology is called an infrastructure BSS (If-BSS). No peer-to-peer direct communication between the stations is allowed in the infrastructure mode. With this topology, networking with a wired network is possible by using an AP.

Figure 13-2 shows the concept of a distributed system (DS), which is used to connect two or more BSSs to form an extended services set (ESS). In this respect, it is like a backbone network for WLANs. It is used to extend the coverage of BSSs in large geographical areas. The ESS is not specified by standards bodies and is a vendor-dependent implementation.

13.2   PLMN-WLAN Roaming

WLAN hot spots are extremely popular among users for high-speed Internet access. They also allow business users to access their corporate networks using secure VPNs, while on the move. In order to use a WLAN hot spot, one must subscribe to the services of a WLAN provider. WLAN operators are trying to increase their user base by offering users flexibility in terms of subscription, security, billing, and charging. At present, WLAN as a stand-alone service is not really making a lot of profits for the service providers.

Wireless mobile networks, on the other hand, possess a huge subscriber base and coverage. With national and international roaming in place, mobile operators may rightfully boast global coverage. However, when it comes to satisfying mobile data needs, mobile networks are slow. GPRS and 3G networks offer significantly higher data rates but still are limited in terms of coverage and speed.

Convergence of WLAN islands and mobile networks is a potential solution to provide a much better mobile data experience to users. To meet the mobile data needs beyond GPRS/3G, users expect seamless roaming between the mobile networks and other access technologies such as WLAN. One of the critical requirements of seamless roaming is transparent end-user authentication and security across different access technologies. Most users prefer one subscription to access mobile services anywhere, anytime, and one bill. One of the reasons behind the success of GSM roaming is its ease of use. A roamer simply switches ON the phone in a foreign network to get the mobile services without bothering with subscription and network specific procedures etc. The same needs to be implemented in the case of roaming between WLAN hot spots and mobile networks. At present, the mobile networks support SIM-based security, authentication, and registration, while WLANs support a mechanism based on user password and access code. The security adapted by the GSM/UMTS network has been proven for many years now. The same cannot be said for the simple security mechanism adopted by WLANs, which are vulnerable to hacking, spoofing, and spam.

13.3   WLAN-PLMN Interworking

Recently, a few wireless operators, in cooperation with WLAN providers
and NEMs, have concluded trials to successfully demonstrate seamless
roaming. 3GPP also released a series of specifications on WLAN and
PLMN interworking in late 2004. Figure 13-3 illustrates WLAN networks
and interworking with the PLMN in a simplified form.

13.3.1   WLAN UE

WLAN UE is user equipment such as a laptop computer or PDA equipped
with a SIM/USIM card or UICC (universal integrated circuit card). The
UICC is a physically secured device that can be inserted and removed from
the terminal equipment. It contains one or more applications, one of which
is always USIM. WLAN UE may support WLAN-only access or WLAN and
3GPP radio access.

The functions of a WLAN-only UE include:

■ Associating with an interworking WLAN (I-WLAN). An I-WLAN is a
WLAN that interworks with a 3GPP System.

■ WLAN access authentication using the extensible authentication
protocol (EAP).

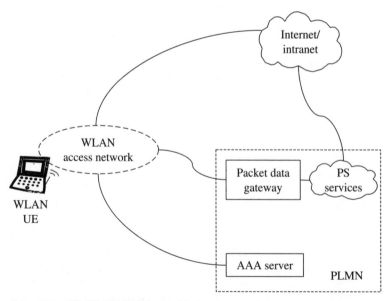

Figure 13-3   WLAN-PLMN interworking.

- Constructing a network access identifier (NAI). The NAI is used to identify the WLAN UE to the AAA server in EAP messages. NAI is derived from the IMSI.
- Obtaining a local IP address.

In addition to above functions, the following functions are required to support a WLAN 3GPP IP access-enabled WLAN UE.

- Constructing a WLAN access point name (W-APN) to identify the external IP network.
- Initiating procedures to resolve W-APN to a packet data gateway (PDG) and establishing a secure tunnel to PDG.
- Obtaining a remote IP address.

In case of roaming, WLAN UE also performs VPLMN selection.

13.3.2 AAA server

The AAA server supports authentication, authorization, and accounting (AAA) for a PLMN subscriber using WLAN access provided by a partner WLAN operator. The AAA server resides in the 3GPP-compliant PLMN, as shown in Figure 13-3. The AAA server communicates with the HLR/HSS to retrieve necessary authentication and subscription information. It communicates with the WLAN access network (WLAN AN) to authorize or deny a PLMN subscriber access to WLAN. Subsequent changes in user profiles (if any) are also communicated to the WLAN-AN.

In cases where the user wishes to use packet switched data services by the HPLMN or the VPLMN, using WLAN access, the AAA server, in addition to its main functions communicates with the packet data gateway (PDG) to pass the service authorization information.

AAA proxy functions may collocate with an AAA server in the same node. AAA proxy is responsible for relaying AAA information between WLAN and the AAA server, filtering, enforcing policies based on the roaming agreement between the WLAN and PLMN service providers, and reporting charging and accounting information for each user to the VPLMN offline charging system.

13.3.3 Packet data gateway

The packet data gateway resides in the user's HPLMN or the selected VPLMN. The PDG is applied only if a user is a PLMN subscriber and wishes to use the packet switched services offered by the PLMN or access Internet/intranet via PLMN. For a roamer, W-APN and the subscription determine whether the services will be provided by the

HPLMN or the VPLMN. The PDG contains the routing information for WLAN users accessing PLMN services and is responsible for data transfer between the PDN and the WLAN user. The PDG performs data screening based on defined rules to filter out unsolicited and unauthorized traffic. It is also responsible for generating charging information for the usage.

13.4 Roaming Scenarios

Figure 13-4 shows a detailed interworking model when a roamer in a visited WLAN operator accesses HPLMN-based packet data services. The visited WLAN is a WLAN operator that interworks with one of the VPLMNs having a roaming relationship with the roamer's HPLMN.

Figure 13-5 illustrates the interworking model when a roamer in a visited WLAN operator accesses VPLMN-based packet data services.

In both the cases, the HPLMN is responsible for the access control. The AAA proxy relays access control signaling and control information to the HPLMN's AAA server.

The WLAN access gateway (WAG) is responsible for enforcing routing of packets through the PDG. In the roaming case, it resides in the VPLMN.

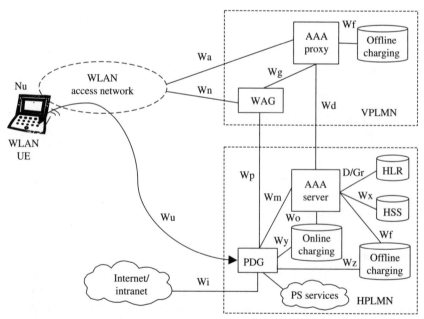

Figure 13-4 PS service access when a roamer in a visited WLAN operator accesses HPLMN-based packet data services.

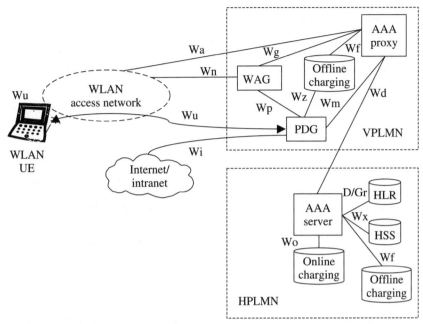

Figure 13-5 PS service access when a roamer in a visited WLAN operator accesses VPLMN-based packet data services.

The HLR and the home subscriber server (HSS) reside in the HPLMN. These entities contain the authentication and subscription data.

As shown in Figure 13.5, both online and offline charging are supported. The system also provides capability to control a specific ongoing WLAN access session for online charging purposes.

The reference points shown in Figures 13.4 and 13.5 and their purpose are shown in Table 13-1

Bibliography

3GPP TS 29.161. Interworking between the Public Land Mobile Network (PLMN) supporting packet based services with Wireless Local Loop (WLAN) and Packet Data Network (PDN).

3GPP TS 23.234, 3GPP system to Wireless Local Area Network (WLAN) Interworking System description.

3GPP TS 29.234, 3GPP system to Wireless Local Area Network (WLAN) Interworking System; Stage 3.

3GPP TS 32.252, Wireless Local Area Network (WLAN) charging.

3GPP TS 33.234, Wireless Local Area Network (WLAN) Security.

TABLE 13-1   Reference Points

Reference point	Description
Wa	Wa is the reference point between WLAN access Network and the 3GPP AAA proxy (roaming case) or AAA Server (non roaming case). The protocols over Wa transport authentication, authorization and charging-related information in a secure manner.
Wd	Wd is the reference point between 3GPP AAA Server and 3GPP AAA Proxy. The protocols over Wd interface transport authentication, authorization and related information in a secure manner.
Wf	Wf is the reference point between 3GPP AAA Server/Proxy and 3GPP Offline Charging System. The protocols over Wf transport/forward offline charging information towards 3GPP operator's Offline Charging System located in the visited network or home network where the subscriber is residing.
Wg	Wg is the reference point between 3GPP AAA Server/Proxy and the WAG. The protocols over Wg provide information needed by the WAG to perform policy enforcement functions for authorised users. It is also used to transport per-tunnel based charging information from the WAG to the AAA Proxy for roaming scenario.
Wi	Wi is the reference point between the Packet Data Gateway and a packet data network. The packet data network may be an operator external public or private packet data network or an intra operator packet data network, e.g. the entry point of IMS, RADIUS Accounting or Authentication, DHCP. This is similar to Gi interface in GPRS network.
Wm	Wm is the reference point between 3GPP AAA Server/3GPP AAA Proxy and Packet Data Gateway. AAA Server uses this interface to retrieve tunneling attributes and WLAN UE's IP configuration parameters from/via Packet Data Gateway.
Wn	Wn is the reference point between the WLAN Access Network and the WAG. This interface is used to force traffic on a WLAN UE initiated tunnel to travel via the WAG.
Wp	This is the reference point between the WAG and PDG.
Wo	Wo is a reference point used by a 3GPP AAA Server to communicate with 3GPP Online Charging System (OCS). The protocols over Wo interface transport online charging related information so as to perform credit control for the online charged subscriber.
Wu	Wu is the reference point between the WLAN UE and the Packet Data Gateway. It represents the WLAN UE-initiated tunnel between the WLAN UE and the Packet Data Gateway. Transport for the Wu reference point protocol is provided by the Ww, Wn and Wp reference points, which ensure that the data are routed via the WLAN Access Gateway where routing enforcement is applied.
Ww	Ww is the reference point between the WLAN UE and the WLAN Access Network.
Wx	Wx is the reference point between 3GPP AAA Server and HSS. This is used to transport access and authentication related information.
Wy	Wy is the reference point between PDG and On line charging system. It is used to transport online charging related information about WLAN 3GPP IP Access so as to perform credit control for he online charged subscriber.
Wz	Wz is the reference point between PDG and Off line Charging System. It is used to transport offline charging related information about WLAN 3GPP IP Access.

Index